T0349829

FROM
MICRO
TO
MACRO
Adventures of a Wandering Physicist

FROM

MICRO

TO

MACRO

Adventures of a Wandering Physicist

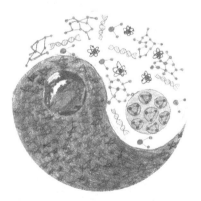

Vlatko Vedral

Oxford University, UK & NUS, Singapore

 **World Scientific**

EW JERSEY · LONDON · SINGAPORE · BEIJING · SHANGHAI · HONG KONG · TAIPEI · CHENNAI · TOKYO

Published by

World Scientific Publishing Co. Pte. Ltd.
5 Toh Tuck Link, Singapore 596224
USA office: 27 Warren Street, Suite 401-402, Hackensack, NJ 07601
UK office: 57 Shelton Street, Covent Garden, London WC2H 9HE

Library of Congress Cataloging-in-Publication Data
Names: Vedral, Vlatko, author.
Title: From micro to macro : adventures of a wandering physicist / Vlatko Vedral
    (University of Oxford, United Kingdom, National University of Singapore, Singapore).
Description: Singapore ; Hackensack, NJ : World Scientific, [2018] |
    Includes bibliographical references and index.
Identifiers: LCCN 2017035579| ISBN 9789813229518 (hardcover ; alk. paper) |
    ISBN 9813229519 (hardcover ; alk. paper) | ISBN 9789813231405 (pbk. ; alk. paper) |
    ISBN 9813231408 (pbk. ; alk. paper)
Subjects: LCSH: Physics--Miscellanea. | Science--Miscellanea.
Classification: LCC QC75 .V43 2018 | DDC 530--dc23
LC record available at https://lccn.loc.gov/2017035579

British Library Cataloguing-in-Publication Data
A catalogue record for this book is available from the British Library.

Copyright © 2018 by World Scientific Publishing Co. Pte. Ltd.

All rights reserved. This book, or parts thereof, may not be reproduced in any form or by any means, electronic or mechanical, including photocopying, recording or any information storage and retrieval system now known or to be invented, without written permission from the publisher.

For photocopying of material in this volume, please pay a copying fee through the Copyright Clearance Center, Inc., 222 Rosewood Drive, Danvers, MA 01923, USA. In this case permission to photocopy is not required from the publisher.

For any available supplementary material, please visit
http://www.worldscientific.com/worldscibooks/10.1142/10696#t=suppl

Printed in Singapore

To my children
Mikey, Mia and Leo — We are the Road Crew

CONTENTS

CONTENTS

PROLOGUE
The Point of It All

This book was born on a sticky, humid night in Beijing, but its conception took place a few years earlier on a very different evening. An Oxford dinner party, Hertford College. Think mahogany, suits, free-flowing laughter and even freer-flowing wine.

Will Hutton was our host with the most — a political economist, writer and television presenter who is deeply concerned with the wealth and education gaps that are ever-widening in the developed world. Will's political instincts are left and liberal, his concerns about the future very much in tune with a majority of the young western generation.

Will is a man of many hats, and this particular night donned the hat of college principal. He is a rare breed of human with impeccable leadership skills subtly underpinned by a friendly and thoughtful charm. Everyone gets on with Will. I challenge you not to.

The dinner was his idea: a gathering of twenty people and a lot of hats — business figures, media people, natural and social scientists. The discussion banner we were gathered under was 'Disruptive Technologies'. A major concern of Will's is that the rising technological tide lifts only

1

a minority of boats and threatens to sink the majority of others. New technologies, while empowering a large fraction of population, also tend to focus the wealth they generate into a much smaller minority. And this, apart from being unjust, is possibly the best recipe for creating social unrest.

Mind you, I say all of this with post-dinner hindsight. This was the first time I had heard the phrase disruptive technology (it took me a couple of minutes to figure out that the drums which my thirteen-year-old son is so fond of playing is not one of them). I am not much of a geek (though some of my friends may disagree) and I am certainly not an economist.

My only possible connection to the rest of the participants at Will's dinner, or so I thought, was my research into quantum information and its biggest potential application — the world's smallest and fastest computer, known as the quantum computer.

Quantum computers, once they are made (notice I am not saying 'if'), will most likely count as disruptive. By disruptive, I mean that quantum technology would lead to a completely new set of problems. It will solve some big existing problems of course, but as with any new technology, it would in turn generate fresh problems of its own. And these new problems, most of which we cannot even anticipate at this stage, would likely require society to readjust in a big way (which would be rather disruptive).

As a physicist, I am more interested in the physics (obviously) — the *why* more than how these new technologies work. I am certainly far from tech-savvy. I must admit a lack of interest in following new business or media trends and typically for a guy from Generation X, my children (all Millennials) have a far greater command of their mobile devices than me. 'Dad, you still have a *Blackberry*? What's *that* about?' they cry, as if I'm Ronnie Corbett in his fruit and veg shop. ('Do you want your Blackberry on Orange?' he asks, holding up the two pieces of fruit.)

But at this dinner, I was really a bit of an outsider. Fluid and enjoyable, the conversation meandered between various topics, from food shortage to global warming to popularisation of science and politics. These are all big and complicated topics, especially for a quantum physicist. We reside largely in the micro domain, and these are certainly macro topics. All I really contributed was my well-versed thoughts about the importance of communicating science to as wide an audience as possible — in this age of technology, it is essential to understand science if we want to know our world and our future. But otherwise, I was lost for words.

Then at some point, Will turned to me. 'Vlatko. What do you think is our next big challenge?' The room went quiet.

What on earth can a run-of-the-mill quantum physicist tell people who are on the battle frontlines of these big global issues? Though I was surprised at being asked, I got what he meant by the question. He wanted me to stick my neck out and predict the next big technologically disruptive trend. This is, of course, very hard for anyone to do.

The words left my mouth before I really gave them permission to. 'Our biggest challenge', I said, 'is bridging the gap between the micro and the macro.'

I had barely finished the statement when others from around the table murmured their agreements. I was rather disappointed, actually, not to have been able to expand on what exactly that means to a quantum physicist, to talk about bizarre experiments with names like 'Schrödinger's cats' and 'Heisenberg's cuts', to wax lyrical about quantum computers. The others got in there first.

The lady opposite me, a chemist, spoke first, acid green eyes beneath long fair hair. 'This is exactly my problem as a chemist. On the micro level, I know enough to *design* drugs but when it comes to understanding the bigger picture, complex molecular reactions on a body — my computer crashes!'

'It's worse in biology — everyone knows that', a large man, with a particularly shiny bald head, said with a wave of his hand. 'Micro to macro has always been our biggest concern. We understand simple organisms of course, but when we try to zoom out — how did living matter decide to be alive?'

The man next to me, an economist with a booming voice and scarlet tie let out a laugh. 'What about me? I'm asked by the government when the next stock market crash will be. I can explain individual financial transactions until the cows come home — but predicting the large scale economic trends? It's a nightmare.'

'Well', a Japanese lady with serious eyes spoke next, as I'm wondering when I'm going to get a word in edgeways. 'If we're talking about understanding macro human behaviour, try being asked to explain why communist revolutions happened. Why big groups act the way they do. I'd go as far as saying your problems are all small parts of mine!'

A consensus was emerging that I had hit the nail on the head. Is it really that the micro to macro gap is the biggest problem in such a diverse set of human endeavours? The feeling that others in a debate agree with you when it comes to a matter of opinion is unusual for a physicist, especially as I had been winging it. (Which of course, a physicist would normally never dream of doing.)

'Allow me to add something,' I said. 'None of the sciences you work in deal in matter as small as we quantum physicists do. We are working on the smallest scale possible. Dare I suggest — that if we were able to bridge our gap between micro and macro — then your gaps may no longer be there!'

For a moment, there was quiet, and I wondered whether I had just hypothetically solved the world's problems. Then the laughter started. 'Solving all our problems!' 'Quantum physics saves the day!' I kept quiet and joined in the laughter. But I started thinking. And kept thinking. And then decided to write this book.

Again, with the gift of hindsight, it is not surprising that everyone could identify with the micro to macro gaps, for the western world owes a great deal of its success to the fact that our knowledge has become increasingly compartmentalised over the last four centuries. This happened when first the medieval Italians, and then others in northern parts of Europe realised that in order to understand the nitty-gritty of various natural processes we needed to do careful experimentation and model it with some precise mathematics. And the two — experimentation and mathematical model-making — have to work hand-in-hand. But the problem is that during this compartmentalisation, while we certainly gained ground-breaking insights, something else was lost — unity, togetherness, and a sense of the big picture.

Combining model-making with experimentation in this way has become known as the scientific method, and the approach spread like wildfire from the seventeenth century onwards. It triggered the industrial revolution in the north of England and the rest is history (of modernity). There is hardly any society on earth today that does not know about, use or at least benefit from the fruits of science, and that probably even includes the indigenous Amazonians and Papua New Guineans.

But science still works in a largely compartmentalised way. In quantum physics, a subject that deals with ridiculously small objects like atoms and molecules, we are frequently unaware of even the most basic details of what other physicists — an astrophysicist for instance — are studying, let alone the chemists and biologists. Even though physics, chemistry and biology are all natural sciences, and all indissolubly linked, they are so divided from one another in practice that there is hardly any communication between any of them.

What my dinner companions had pointed out was that this is not just the case for the natural sciences. In the social sciences, it is exactly the same. Economics, said the man with the booming voice, is linked closely to human psychology, yet economists and social scientists rarely mix. Microeconomics involves financial transfers between a small number of individuals, and usually just two people. Yet macroeconomics is about the policies that whole countries are, or ought to be, pursuing.

There is an example in economics where one particular gap has been bridged — leading to increased understanding. Microeconomics is now very much modelled on physics, to the extent that many PhD students of physics and mathematics switch their focus to economics, thereby usually greatly improving their personal finances and, some would say, increasing the chances of getting a Nobel Prize.

Microeconomists use rigorous mathematical models, which are able to give us definitive predictions about what kind of preferences people would have in different situations. And making predictions is crucial for something to be considered a proper science. Unless the predictions we make can be tested and verified (even in principle, but better if this is also true in practice) it is hard to call something a hard scientific fact.

Why and how people make choices is exactly what economics is all about. Fascinatingly (well, to physicists at least), microeconomists also do experiments. Yet there are frequently discrepancies between what the economic models predict and how real people behave. To a physicist, this is to be expected. And if it is clear to a physicist, it is even clearer to psychologists who deal with human idiosyncrasies on a regular basis.

People are different from atoms. They have feelings, emotions (even physicists), and all sorts of charming individual quirks. On top of that complexity, we also think that we make our own decisions and that we have the free will — for lack of a better phrase — to 'change our mind'. Atoms, and other objects of physics, do not. They blindly follow whatever exact prescription the laws of nature give them.

And this human complexity when compared to the atomic world is a problem for economists even before we get to macroeconomics! Can the two ever work together? Can microeconomics (and the mathematics it uses) and the social sciences (human behaviour) ever be integrated?

Let me tell you a little story, which shows that they indeed can. The little story is also a true one, about a friend of mine whose identity I shall keep hidden for the sake of discretion (not that he would mind).

This was his dilemma: he was to be married, and he had a certain number of ladies to talk to before making his decision. To make matters complicated (as if getting married is not complicated enough) he needed to do it using the following model:

So, on day one, you have lunch with the first randomly-chosen lady from a list of 'candidates' (for want of a better word), but already by the evening you have to reply to her with a definite 'yes' or 'no'. Once you have said no, you cannot return to this person and change your mind (thankfully, for the pride of this poor woman).

Then on day two, you repeat the procedure and have lunch with the second lady on the list, to whom you also have to give a definitive answer by that evening. And so on. The question is, at what point is the optimal time to say 'yes'? Say you have seen four ladies, said no to them, but the fifth one has impressed you more than the previous four. Should you propose to her? What if you have still got 10 more ladies to talk to? Is it premature to propose to the fifth out of 15 candidates?

Hang on though — why would the procedure be such that you have to give the answer to one lady before you have met the next one? In my friend's case, this was his cultural norm. The family of the lady you have just had lunch with expects the answer by dinner time on the same day, otherwise you are being disrespectful to the family. And that counts a lot against you in your future dealings — word gets round — making possible future wife candidates less likely to apply.

You may think this is terribly unromantic, but most of the world still does something like this in order to hook up. Those of us in the west are actually exceptions, and at that, only in the last 50 years; before that, most of the world followed the model of arranged marriages.

Interestingly, this particular problem of tying the knot has an optimum solution. Obviously, the more ladies you have on your initial list, the more of them you should meet before committing. But there is a precise answer to this question when the sample of ladies is large. The solution says, meet the first n divided by e ladies (n being the total number of candidates and e being the mysterious and all-pervading Euler number equal to about 2.71). After that, propose to the first lady that is better than all the previous ones you have seen.

You can prove this result using some simple algebra which I would not bore you with here, but this optimal solution is creepily close to our human intuition. How many ladies would you meet before committing? A half, was the general answer when I asked different friends and colleagues this question. Now, the solution of $1/e$ is somewhere between a third and a half, so their intuition was not too far away.

Let me get a bit macro here for a minute. As a politician you might not care much about how individual couples meet and agree to get married. The nitty-gritty of the hook up process is not really something we think politicians should interfere with (it is none of their business, really, though they frequently want to make it so). But politicians certainly do care a lot about the overall marriage trends in their country. They are aware that stable marriages are more likely to lead to more stable, better-educated and well-rounded children, whose contribution to the society will, on the whole, be more positive (excuse my generalising).

It would be very powerful to take the model I just described and derive larger societal trends. Imagine a politician trying to avert the 'ageing time bomb' — the fact that in developed countries the death rate exceeds the birth rate, meaning sooner or later we will have one adult having to work for his young family in addition to supporting one retired person. Having access to a good microscopic model telling him how to predict how many married couples and how many kids they will have on average in 50 years' time could help a lot in informing policy.

My friend and his ladies? He had seven ladies in total on his list as potential spouses (the lucky seven shortlisted, down from some 70 plus candidates). He proposed to the third one, which is pretty close to the optimum solution of $7/e$, which equals 2.6. (For fear of being seen as unromantic, let me assure you she is a wonderful woman, who I'm sure would not be too happy being known to you only as a figure in an equation.)

I know, I know... the marriage model I presented is very simplified. It does not take into account a lot of other things, free choice and our natures and various other human imperfections. But it illustrates the point I'm trying to make. We often do not know how to predict the social macro trends even if we understand the micro ones — a model like this, although simplistic, can teach us a great deal. Definitive predictions can be more easily falsified (and therefore improved) by practice. And this is the key: keep refining until the predicted outcome is indistinguishable from experiments.

And if we *do not* try to bridge the gaps between micro and macro? The result: more separatism, less unity, less understanding, and a lot of time wasted. We do not have endless time, and we do not have endless other resources, such as energy or even money. On an environmental level, you can see why it is in the interest of each individual country to continue polluting (increasing industry, infrastructure and growing economically). But you can also see that if everyone continues with this behaviour, we will all soon be much worse off. Can microeconomics and social science offer us a glimpse of how to address macro environmental issues?

Oh, I do love an anecdote, so please excuse one more to illustrate an event in history that will continue to repeat itself unless we learn from the trouble gaps can cause. Okay — two more. The first is the well-known prisoner's dilemma. So, police interrogate two prisoners, separately and in different rooms, without allowing them to communicate. To make things simple, each of the prisoners has two options: to cooperate (meaning, he keeps quiet and does not admit anything to the police, thereby cooperating with his partner in crime), or defect (meaning that he admits crimes to the police, thereby grassing on his friend.)

The temptation to defect is, of course, that your sentence is reduced (that, after all, is how the police entice you to defect). However, if both prisoners cooperate, the sentence would also be shorter. The worse outcome is for both to defect (since then both are proven guilty — it does not matter that they admitted it). But when we analyse the game rationally, this is what they ought to end up doing. Here is why...

Imagine you are one of the prisoners being interrogated in a separate cell from your partner in crime. You are thinking, if I defect, my sentence is reduced (though my partner's increases if he keeps quiet). But if my partner defects, then it is clearly stupid of me to cooperate. I should defect. Therefore, whatever my partner does, it is always better for me to defect. Since your partner is thinking the same way (as the game is perfectly symmetric), she also chooses to defect. And this is the worst outcome for both of you — you would be better off both cooperating (but since you are interrogated separately it is hard to reach this agreement).

A many-player prisoners' dilemma is known as the tragedy of commons, taking us back to the environmental domain. Suppose many villagers can use the commons for their sheep to graze. However, there is an optimum number of sheep for the commons to be efficiently maintained. Too many sheep and the commons is destroyed. Too few and the resources are underutilised.

So, no one has the incentive to add more sheep beyond a certain number, right? Wrong. As in the prisoners' dilemma, each villager benefits by not cooperating with others and adding more sheep. The cost of maintenance is borne out by the whole village and the individual therefore does not suffer much as a consequence. But if everyone followed this logic and abused the grazing commons, then of course, the state of affairs would soon reach a point beyond repair.

We no longer have the village commons (maybe the situation described above is the reason?), but we have plenty of other instances of similar behaviour — the individual acting how it pleases without regard for the best interests of the collective. If the micro can explain the macro, if more gaps can be bridged, we can perhaps hope to save our planet, and our human race, for a little longer than they are currently looking to last.

My own micro to macro research interest lies in the area of applying quantum physics to larger and larger objects. Quantum physics is faultless when it comes to describing small particles, such as atoms and photons (particles of light). However, it is clear that large objects, such as tennis balls, obey different laws of physics to quantum. The laws for tennis balls are entirely due to Newton. Force = mass × acceleration. Somewhere between an atom and a tennis ball there might be a path that leads from the micro to the macro domain. My research is about probing this particular transition.

I am a physicist and I tend to try to understand the whole natural world around me through the eyes of a physicist. In this, I am by no means an exception. Physics itself is increasingly being taken outside of its own domain of studying interactions of matter and energy in order to be applied to systems of greater complexity. But why? To what end?

Take for instance, systems studied by chemistry and biology. It is generally accepted that the laws of quantum physics underpin the laws of chemistry, such as how molecules are formed by gluing atoms together. As far as biology is concerned, physics has even claimed a sub-field therein: bio-physics itself is a growing discipline of huge importance with applications ranging from drug testing to bio-engineering.

This makes sense on the larger time-scale: we think of the laws of physics originating somewhere at the same time as the universe, which was 13.7 billion years ago. And chemistry only started to matter at a much later stage when stars were formed and heavier atoms were created in them. Biology kicks in even later with the evolution of life and this is estimated to be 4 billion years ago (at least as far as earth is concerned).

But the impact of physics is felt far beyond the natural sciences. Economists and other social scientists are now starting to think in a mathematically quantitative mode akin to the way we build models in physics. Economics starts with humans, but some of the mathematics that is used to describe human transaction is also extremely potent in biology. The little stories I described earlier are typical in micro economics. Social sciences are also starting to employ sophisticated experimental techniques whose origin firmly lies within the methods pioneered by physics. Gaps are starting to close, and we are all benefiting.

So why this book? What do I hope to share? A journey is what I have in mind: a journey through space and time, peering in at the micro to macro gaps of different disciplines, seeing where gaps have been bridged thus far and which ones remain. We will fly between different sciences: around Oxford for a portion of physics; to Beijing and the Great Wall for a bigger picture of chemistry; over to Singapore for a trip through biology; to a potential Metallica gig in Dubai for a blast of economics; finishing in a Belgian theatre to applaud our way through the social sciences. Everywhere I went on my travels, the question of gaps and reduction was never far from my mind, along with the biggest question: Are gaps part and parcel of the scientific method, or can they all disappear one day?

With every scientific discovery, a new gap is closed. The gaps themselves have different origins and causes. Some of them exist within the micro domain itself, some in the macro domain, while the rest are to do with the gap between the two. We will explore mathematical, technological and philosophical roots of the gaps, as we question why they are still open. Perhaps the gaps left are simply what we have not yet uncovered. The micro to macro gap will be our focus. And if it is possible to close the biggest gap in physics (between the micro quantum and the macro gravitational description), would this in turn close the gaps in other sciences? For the purpose of this book, I will refer to this as the 'Great Reduction', the hypothetical event in which all micro to macro gaps could be closed.

It was with a slight bit of hesitation that I began to write this book, for I am very aware that I am a simple physicist, with no real authority to talk on subjects outside of physics. I am sure I will over-simplify things along the way and possibly cause offence to those of other scientific disciplines, and for that, I sincerely apologise. I hope you will be able to see beyond my ignorance to the bigger picture, as I attempt to weave a complex tapestry.

I also speak of reductionism in a narrower sense, which is a subject so broad it is a wonder its different variations can be categorised using a single word. There are enough types of reductionism to write a book of more than this length about, and I am aware that I speak merely of one type of reductionism more or less. There are many different views as to what it would mean to reduce one theory, or one set of phenomena to another one. Again, please accept my apologies for my own reductionism of reductionism.

We live in exciting times, we all agreed, at our Oxford dinner party. We are in the middle of a scientific revolution. The natural and social scientists had plenty to talk about on this topic. Before this revolution began in the second half of the last century, science was separated brutally, each discipline understood in isolation. Post-World War Two, our understanding finally started being reduced to the fundamental quantitative laws, just like the laws of physics. But we still have some way to go.

Why try to understand the nature of these gaps? Why speculate as to whether the Great Reduction will ever happen? Yes, of course, our intellects are deeply satisfied by simplifying laws — namely, by finding a smaller number of them — and thereby reducing the complexity of the underlying explanations. 'The intense pleasure I have received from this discovery can never be told in words', Kepler tells us upon discovering his laws of motion which explained apparently complex planetary motion in terms of very simple geometrical facts. But I believe there is much more at stake.

I believe that any society that achieves this kind of deep and complete understanding of reality will become far more superior to other societies, both spiritually and technologically. The former will lead to a great well-being of humankind while the latter might well be our only exit strategy from this overpopulated, resource-depleted and environmentally-polluted planet of ours.

Even more than that, I believe that the very difference between the spiritual and the technological spheres would itself vanish if this Great Reduction was achieved. Religions and philosophies have been constructed with the principal aim of giving us a complete and coherent picture of the universe. They give us safety and comfort. I believe the Great Reduction could do the same. The longing for connection and completion underpins humanity's deepest desires and motivations. If science is to deliver this overdue prescription, it could only be through closing the gaps that prevent the unity that might very well save us.

CHAPTER I

PHYSICS AND ITS TROUBLESOME GAP

If you want to understand how a car works, you have to get your hands dirty. You have to get up close and personal with its parts (NPI), before you can really understand the beast itself. The micro laws of each part give rise to the macro ones. And this, intuitively, is how it should be. The whole is the sum of its parts.

In physics, we are working on a much smaller scale than car parts, but the logic is the same. We are, of course, talking atoms (and in the context of this book the motion of atoms, rather than types of atoms, which many more books could be written about). The first genius to explain how the individual motions of atoms can be combined to explain the collective behaviour of stupendous numbers of them was Austrian physicist Ludwig Boltzmann. Common sense might tell us that it is impossible to say anything accurate about a gazillion atoms since each of them moves in its own particular and, to a large degree, random way. But randomness is the thing that helps us here, in walking the first line in physics between the micro and the macro.

Let me give you an idea of just how small we're talking. Atoms are measured in nanometres; a nanometre is the size of 10 atoms lined up next to each other. Mathematically, it is a billionth of a meter. Humanly, it is about the amount your hair (or beard) grows by in a *second*. In other words, your hair grows by some 10 atoms every second. If you have just had a bad haircut and all you can do is sit and wait for it to grow back, you will have to "hermit-ise" yourself for about a month until your hair gets a few centimetres longer. And that few centimetres is about 100 million atoms sitting next to each other in a straight line.

By comparison, the size of the visible universe is roughly 1000000000000000000000000000 bigger (that is 10 to the power of 27 times bigger). You will be very grey by the time your hair grows that long. In terms of size, the micro to macro transition we are discussing involves traversing this vast space and understanding all the phenomena in between. What could go wrong?

Nowadays, we know that atoms exist because we can actually see them under a microscope (albeit a very special, Nobel Prize-winning microscope), but this was not the case a hundred years ago. Then we only had indirect evidence, and this evidence is one of the key links between some micro and some macro aspects of the physical world. If we could not see them, how could we possibly have known that they existed?

Let us go back to me for a minute. I remember the first time the idea of atoms popped into my head. I was a young spotty teenager taking a flight with my school on a trip to the stunningly beautiful Croatian seaside. I was looking out of the window when it occurred to me that the aeroplane engine must only be half the story behind how aeroplanes could fly. I mean, the engines push the plane forward, but what gives it the upward thrust to lift off? It is obvious with rockets — their engines blast off fuel downwards giving the rocket an upward push (thanks to Newton's 'action and reaction'). But not with planes, where the weight of the air above the plane makes it even harder to take off.

The twist in the story is that when the aeroplane is moving, the upward pressure of atoms going below the wings is greater than the downward pressure of the atoms going above the wings. And there we get the upward thrust, the motion of atoms and the pressure they exert when they hit the wings. *If there were no atoms, the plane could not fly.* This is why planes do not go above 35 thousand feet — there is hardly any air there. And this is what struck me like a bolt of intellectual lightning. The fact that we have flying planes is indirect proof of the existence of atoms (I say 'indirect' because direct proof would constitute directly seeing or experiencing them physically). An unusual train of thought for a teenager, but there we go, this is what floated my boat. Yes, before you ask, I was single for a lot of my teenage years.

Not getting the balance right between upward and downward air pressure is actually the cause of more than half of all aeroplane crashes in the last 20 years. If the aeroplane does not fly straight, but at an angle, then at some point it loses the upward thrust, leading to what is known as the 'aerodynamic stall'. Getting out of the stall — which leads to rapid descent of the plane — requires a careful pilot manoeuvre. But sometimes, it is too late for the pilot to do anything. This is what happened to Air France Flight 447, a scheduled passenger flight from Rio de Janeiro to Paris which crashed on 1 June 2009, killing all passengers and crew, 228 in total.

The Wright brothers flew the first plane in 1903. So already we must have known (though we did not quite *know* we knew it) that atoms existed. Einstein's paper in 1905, only two years after the Wright brothers' flight, was final accepted proof of their existence. Coincidence? He did not use aeroplanes, though the spirit of his logic is remarkably similar. He used particles of dust that execute a random motion we now call Brownian, after a biologist who first observed this motion in the eighteen century. A speck of dust suspended in a tube will move erratically as if it was pushed around by an invisible drunken ghost.

Einstein hypothesised that the reason for this is that the speck was bombarded — not by a drunken ghost (for why would he waste his time pushing specks of dust around?) but by a lot of small objects (atoms) which move about in a random way. He went even further. By measuring the precise details of the motion, he could deduce the number of atoms in the tube where the dust moved about. And he got what is known as the Avogadro's number, roughly the number of atoms in two pints of water — a one followed by twenty-six zeroes (we like our big numbers in physics).

In the same year, Einstein made another micro to macro connection. This one led Einstein to conclude that if a gas consists of atoms, light also must consist of them. The atoms of light are called photons. To reach this conclusion, he turned around our usual way of using the micro to explain the macro. He started with the macro.

The equation of light in a box (using thermodynamics) gives the same link between its entropy (disorder) and volume as the corresponding equation for a gas comprised of a great number of atoms. Einstein therefore argued that since the behaviour of light and ordinary gas is the same at the macro level, it must have the same underlying micro cause — a powerful reasoning that we will encounter over and over again in various sciences. Ergo, Einstein says, there are atoms of light. Being Einstein, he did not stop there. He went on to propose an experiment that would provide more direct evidence of photons — this was the photoelectric effect for which he was awarded the Nobel Prize in physics in 1921.

Going back even further than Einstein is a rather ingenious discovery made by Daniel Bernoulli. It is particularly ingenious given that scientists doubted the existence of atoms until Einstein's prolific paper-producing year of 1905. Bernoulli came from one of those miraculous families that produce consistently brilliant children, with Daniel, Jacob, Johann and Nicolaus all making important contributions in science and maths. What Daniel did was connect macro gas behaviour with the micro motion of atoms. This is an important gap-filler if ever there was one. It led to the steam engine.

I remember the exact moment I understood this connection. In high school, aged 16, I finally had my prayers answered and was graced with an inspirational physics teacher, Mrs Bojana Nikić. It was the first and last time in school that I had a great physics teacher — one of the reasons I knew I was meant to be a physicist was that I loved it despite being taught it badly.

Mrs Nikić said to us: 'I will now use Newton's laws to derive a thermodynamical equation of state for a gas.' Most of the kids in our class blinked up at her as if she was talking Yiddish. But not me. This sounded like magic to me. (Yep, still single.)

The equation of state links three properties into one formula — the pressure of the gas, its temperature and its volume. That is the one where the pressure and volume are proportional to temperature. If we increase the volume at the same temperature, we decrease the pressure. Larger volume means that the gas is less dense and molecules hit the walls less frequently — since they have more distance to travel. Hence, the pressure, which is nothing but the force exerted by molecules as they hit the walls, drops. The thermodynamical equation describes the macroscopic properties of a gas and does not at all care about the fact that the underlying microscopic behaviour is based on atoms and molecules.

This deduction of the equation of state from the motion of atoms is the first significant Reduction in physics, and we owe it all to you, Daniel.

Bernoulli imagined that the pressure of a gas is due to its being composed of tiny atoms moving about and hitting the walls of the container the gas is confined to. Then, using Newton's laws, he was able to argue that the pressure these atoms exert when bouncing on the walls is actually proportional to the number of atoms, their mass and velocity (squared). Subsequently, he used a bit of statistics, which was needed since the atoms are moving about in a completely erratic way — there is an equal chance of finding an atom moving in any direction. After adding this to his logic, Bernoulli showed that the pressure of the gas is actually proportional to the temperature (which is just the energy of the atoms, namely mass times the velocity squared) and inversely proportional to the volume.

And so, the equation of state tells us that any two gases with the same volume and pressure and temperature, actually also contain *the same number of atoms*. This itself is fascinating, but to think that this comes down to Newton's law of force = mass/acceleration blew my 16-year-old mind, and still does today. How can one not love physics? It shows us beautifully that seemingly unrelated facts are actually consequences of one and the same rule.

Connecting the macro gas behaviour with the micro motion of atoms is no doubt intellectually pleasing. It is always beautiful when a more complex phenomenon is reduced to a simpler underlying one. But this has technological advantages too, and working at this macro level, explains how the steam engine was made. We heat it up, it expands and by expanding, it can do work. But, if we had access to the micro side of this, the atoms, we could actually use it more efficiently and encode far more information into it. This is why Richard Feynman, Nobel Prize-winning American physicist, wrote a paper called 'There is plenty of room at the bottom' (what a title). For Feynman, this meant that if we venture into the nano and quantum domains, we could do much more technologically speaking. The first Reduction changed science, changed technology, and changed the world we know.

Another one of the first big bridging triumphs of quantum physics was explaining the behaviour of large solid bodies (like sugar cubes, or pieces of chalk or anything that is not alive) using the underlying quantum behaviour of atoms. One of (many) puzzles before quantum physics was this: as we put the solid in a hot environment it starts to become hotter itself. As we ramp up the external temperature, the solid absorbs more heat, thereby increasing its own temperature. The puzzle? Classical physics could not explain why this happened! Classical physics suggests that the heat capacity of a solid (as the quantity is known) should be independent of temperature. In reality, it actually goes up with temperature.

The only way to explain this is to suppose that electrons move as the solid heats up and its atoms jiggle about harder and harder. And the higher the temperature, the more the electrons move and the harder the atoms jiggle, hence the higher heat capacity. Classical physics simply cannot account for this. And the first person to point this out was — you guessed it — Einstein again. At some point, he turned against quantum physics ('God does not play dice', he tells us firmly), but in the early days, he did more to establish it than any other physicist at the time. Needless to say, his early intuition was the correct one — quantum physics has still not encountered any experiment it is not able to explain with great precision. Physicists love to close gaps. That is why our biggest all-time frustration is the one that remains resolutely open.

THE BIG GAP

Enter Einstein once again, this time with his theories of relativity — the special one is another one of his 1905 papers (it was a busy year) while the general theory of relativity was completed in about 1915. Let us zoom out big time from the quantum level of atoms, right out to one of the forces that governs our universe: gravity.

The universe, we have long believed, is governed by four forces; electromagnetism, gravity, and weak and strong forces. Electromagnetism and weak and strong forces all fit nicely into quantum explanations, and no gaps are left there. But not gravity. This is why quantum physics and general relativity (which explains gravity) do not jell. And this is the biggest gap in physics, and in terms of size, the biggest in our whole understanding of reality. Gravity dominates at large distances, and describes the global features of the universe (planets, stars, clusters of stars, galaxies and so on). Quantum physics describes objects on the very smallest scale. Both theories have been very successful in their own domains. But they just cannot be unified. Like champagne and cigars. Both wonderful, life-giving delights, but they will not work together.

Let me briefly explain how and why (the physics, not the consumables). When quantum physics is applied to the theory of electromagnetism, one of the key consequences is that the energy in electromagnetism cannot be continuous, but comes in discrete chunks, called photons. When we use a laser to excite an electron in an atom, this electron typically goes back to the lower energy state at some point — this process is accompanied by the emission of a photon. The photon is the messenger of the electromagnetic force. Weak and strong forces operate inside a nucleus and when quantised, they also lead to particles (the analogues of photons) that mediate their interactions. The quantum theory of electromagnetic, weak and strong forces (and they can all be thought of in a unified way) is known as the Standard Model and is our best account of all physics. Excluding gravity.

Here comes our trouble-maker. Following the other three forces logically, when we apply quantum physics to gravity, one of the outcomes should be a particle called graviton, the quantum messenger of the gravitational force, in the same way that the photon is the messenger of the electromagnetic force. Maybe graviton is emitted, but the biggest problem is that we, er, cannot even measure it, let alone see it. Now, gravitational force is some 10 to the power of 39 times (1 followed by 39 zeros) weaker than the electromagnetic force. The rate of graviton emission is so small compared to the photon emission that it would take us (much) longer than the age of the *universe* (13.7 billion years) to ever observe it. Damn.

Surely by logic of the other forces, gravitons exist even if we cannot see them. But we quite like proof of things when it comes to science. Speculation is powerful and important, but experimentation and subsequent evidence is the glue that holds theory together, which is certainly needed if this gap is to be closed. There is also one other tiny problem. Trying to quantise gravity actually leads to predictions that are ridiculous at the other extreme. Namely, rather than forecasting tiny numbers, they actually lead to infinite quantities. The predictions for emission of gravitons I discussed are actually based on truncating this infinity. However, it is hard to do this in a consistent way. There is therefore a mathematical inconsistency at the root of the problem too.

So, here we are, standing at the edge of the biggest gap in physics, one that has kept many a physicist awake at night over the last hundred years. This is a genuine micro to macro gap since quantum is so far all about small objects while gravity concerns planets, stars and galaxies. Will quantum physics and gravity ever be unified? And if they were — what could be achieved?

Solution One: It could be that although quantum physics and gravity cannot be put together in a mathematically rigorous way, this might not be an issue after all since any possible consequences are just too small to ever observe experimentally. The gap is only really a gap in our observations, and there is not too much point losing sleep over something we would never be able to see anyway.

Of course, my eager mind certainly does not want to stop there. It wants to look at Solution Two: Perhaps this gap can be closed — by looking at everything in a different way.

But, wait a minute. What could be the benefits of this gap being closed? Why bother reading seven pages of my Great Reduction hypothesising? Well. Back to what I predicted in the prologue — that uniting the micro and macro will help us both spiritually and technologically. First, there is the spiritually satisfying feeling of understanding the whole physical universe with one theory only. This will either be a new theory of quantum gravity or a realisation that gravity is a consequence of quantum physics. With either theory, the spiritual benefits will be the same — science will provide the comfortable world-view that can now arguably only be found in religion.

The technological benefits of bridging the quantum gravity gap could also be enormous. The possibility of efficient space travel is becoming more and more important. We might not be able to resolve our environmental problems on earth in time to make it continuously habitable. In that case, it is very important to understand gravity properly since it determines to a high degree how far and how fast we can travel. If quantum effects matter in strongly gravitational systems, then maybe quantum gravity could improve our ability to travel to distant places.

Also, perhaps quantum effects make the universe work differently, and maybe distant parts are actually somehow quantumly connected. This is all speculation, of course, but speculation is exactly where all theories begin. And maybe, just maybe, this one will be crucial for our future.

A POSSIBLE SOLUTION

Many new ideas in physics are introduced out of desperation. Planck called his quantum hypothesis — when he introduced the idea that energy comes in little chunks and is not continuous as classical physics would have us believe — an 'act of desperation'. It was an act of desperation because it did not make sense to Planck, other than the fact that it was the only way he saw in which to match theory with the experimental results. But he did not have any deeper way of arguing for quanta. That had to wait another 25 years, for Born, Heisenberg, Jordan, Schrödinger and Dirac to discover the full formalism of quantum physics. Sometimes, the bigger picture of a new theory is revealed later — but the beginning often starts with a leap of faith. So, if you will allow me to leap...

At the heart of my speculations are thermodynamics and information theory. I will go into each in turn and build up my case for how they could potentially become our Great Reduction. Before that, however, it is time for some fun. Perhaps it was Will Hutton at that Oxford dinner party that gave me the taste for this, but I would like to stick my neck out once again.

I would like to suggest that thermodynamics is so robust precisely because its foundations are information-theoretic in nature. I am so confident about this, that I am willing to bet a good bottle of Dom Perignon that thermodynamics will also survive whatever revolution comes after quantum physics (although, by then we may well have bypassed drinking and I will have to instead owe you a computer-simulated champagne experience rather than the golden elixir itself).

This is not just a blind bet (I am not winging it this time) — I also have some evidence. Of course, I have no idea what (if anything) comes after quantum physics and relativity. But the very fact that they do not get on well is our key indicator that there might be something beyond.

In the absence of experimental discrepancies, people are studying all sorts of post-quantum scenarios theoretically. These theories all have to maintain the same degree of genuine randomness as quantum physics since this is so well established in all quantum experiments. In other words, they are all probabilistic. They also respect 'no-signalling', in the sense that they do not allow faster than light information transmission.

There are other ideas that have survived the onslaught of post-quantum physics, and yes, you have guessed it, they are thermodynamical and information-theoretic in nature. I have studied this extensively with my colleagues Markus Mueller and Oscar Dahlsten. We have found that the link between information and disorder always remains in the new theories. So, these theories are also unlikely to violate the second law of thermodynamics, as the relationship between information and disorder seems to be the key.

Now I'm really going to take a walk on the wild side. I want to make two predictions for the post quantum world, although this time, I do not quite dare to bet more than a bottle of Bollinger. The first is that (drum roll please) … quantum entanglement survives in the post quantum world.

Entanglement is a form of correlation between two or more sub-systems that go beyond anything classically allowed (which is why Einstein called it "spooky action at distance") and it has been confirmed experimentally in many systems and scenarios. But, more importantly for my argument, entanglement is also linked to information theory and thermodynamics. It is quantified by entropy (like disorder) and it changes only in one direction when subsystems are addressed independently (which, in some sense, is equivalent to the disorder increase).

The other prediction I want to make is that the link between an object's entropy and its area, the so-called Holographic principle, will also remain true in the post quantum world. The Holographic principle is a generalisation of the famous Bekenstein-Hawking formula for black hole entropy (which also relates it to its area). This principle has an information flavour; it tells us about information that is 'lost' when an object falls into a black hole. It has a thermodynamic underpinning too, since it relates this lost information to the black hole entropy increase. Some people believe that the entanglement between stuff inside and outside of the black hole is responsible, which is even more proof.

Ok. Prediction time over. Let us get down to business — information and thermodynamics, and how they may be used to hypothesise the Great Reduction. What better place to begin than on the streets of Oxford, the city which has contained many 'acts of desperation', many of which led our scientific thought to where it is today.

INFORMATION

Oxford is the oldest university city I have lived in. A lot of its most beautiful architecture is built out of limestone, which gives it a soft, magical appearance. Finding myself with a bit of free time and itchy feet earlier today, I walked out into the sunshine that had turned Oxford golden. I took a stroll first through University Parks, marvelling at the information I experienced on every level: in the leaves, the bark of a terrier, the chatter of academics. Information is the fundamental building block out of which everything around us is constructed — information, rather than energy and matter, or cheese and crackers, or whatever else. The purpose of my walk was to ponder the question: If *everything* is built from information, then can gravity also be derived from information theory? And what would that mean for Einstein's equations?

My walk led me down Broad Street, and not for the first time, I noticed the sheer number of stern-looking men (albeit immortalised in stone) staring down at me from the facades of different buildings. This small, delicate city has been a hotbed of scientific discovery over the last few decades, and it humbles me to walk in such great footsteps. I end up inside the Museum of the History of Science, where you can walk quite literally down the chandelier-lit stone interior through all the strands of Oxford science, right down to the basement where I found myself in front of the very blackboard that Einstein used to explain his special relativity in a series of Oxford lectures. And it brought me back to my question. What would it mean to Einstein's equations if gravity can be derived from information theory?

Quantum physics naturally lends itself to this view, as the way we describe the world within quantum formalism is already very much information-theoretic. First and foremost, we talk about states of a physical system as 'catalogues of information' — a term that Schrödinger himself (who was in Oxford just a few years after Einstein wrote on that blackboard) coined when he invented quantum physics. These states give us probabilities with which we should expect experiments to yield different results. The physical evolution of the system is then just a rule telling us how this catalogue of bets on the future states of the system should be updated with time (we call this rule for updating bets 'the Schrödinger equation').

This is all fine, but, of course, there is this blackboard. In addition to quantum physics, we have another equally successful physical theory, namely general relativity, that describes gravitational phenomena. And, yet, quantum physics and general relativity have not been unified. In order to understand the world, we seem to need both quantum physics and general relativity. Unifying them is possibly the biggest open problem in physics today. So if I claim that information is fundamental and that quantum physics is naturally a theory of information, what happens with gravity? I look into Einstein's eyes and ask him this question out loud. Luckily no one else is in the basement. 'Is the answer in thermodynamics?' He does not reply.

In 1995, a beautiful paper was written by Ted Jacobson who showed, lo and behold, how to derive Einstein's gravity — from thermodynamics. This is huge. His paper was (and still probably is) considered very controversial, because it might imply that all those physicists trying to quantise gravity are simply wasting their time since, according to Jacobson, gravity is not a fundamental force, but merely a kind of thermodynamic quantum noise. I signed adoption papers for his logic and extended it to suggest, abashedly tongue-in-cheek, that gravity can in fact be derived from information theory (albeit with a little bit of help from quantum entanglement). Of course, there are a lot of holes in the logic, but also — I hold Einstein's gaze — a sliver of possibility.

It is very simple to explain how gravity might arise from entropy. First, we acknowledge the fundamental thermodynamical relationship that entropy times temperature equals heat. Heat itself is nothing but a form of energy, which according to Einstein, equals mass times speed of light squared. The entropy we assume to be proportional to area (the so-called Holographic principle) in other words, proportional to radius squared. The temperature is, according to Paul Davies and Bill Unruh, proportional to acceleration, which in turn is force divided by mass (from Newton's second law). Putting all this together gives us the force equal to the product of masses divided by distance squared, namely — Newton's gravity!

How watertight is this argument? How much of gravity is really just thermodynamics? The derivation involves two crucial relationships we believe to be true, but which, at present, have no experimental evidence to support them. First, the relationship between entropy and area, and then the Unruh-Davies beautiful formula linking acceleration to temperature. Intuitively speaking, we lack experimental evidence precisely because the relationships are all about linking micro to macro.

This might mean that, if Jacobson's logic is correct in spirit if not in all its details, we need to make corrections to Einstein's field equations in order to accommodate possible changes to the Holographic principle and the non-existence of a unique Davies-Unruh temperature. This is obviously exciting. Not only would it confirm that gravity is not a fundamental force, but it would also point to the need to modify Einstein's equations. I am a few centimetres away from Einstein's blackboard. It would take only a slight brush of hand to change what he has written. I back away instead, one heavy-heeled footstep at a time. Much as one would like to speculate about further consequences of all this, it is for now probably best to exercise the scientific method of suspended judgement. Only time indeed can tell what will turn out to be correct.

THERMODYNAMICS

As I step out of the Museum of the History of Science and back into the late afternoon sunshine, let us get to the 'heat' of the matter: thermodynamics. Thermodynamics is the movement of heat from a body of higher temperature to a body of lower temperature. It is the sun warming me through as I walk back along Broad Street. When you sit on a chair that was just occupied by someone else, it feels warm because your temperature is lower, but the heat will soon transfer to your bottom and warm it up. It also, incidentally, explains how everything we can see exists.

Every time physicists face experiments that cannot be explained with existing theories, they have to decide which aspects of these theories to keep and which to throw away. Planck and Einstein, when faced with the inability of classical physics to explain black body radiation, decided to keep the laws of thermodynamics, but threw away the assumption that energy is continuous (which is an integral part of Newtonian mechanics). Similarly, Einstein, when trying to explain the inability of the Michelson and Morley experiments to detect Earth's motion through the ether, kept the Newtonian assumption that the laws of physics should be the same in all reference frames, but he also introduced the invariance of the speed of light in different reference frames (a fact that is naturally encoded into Maxwell's theory of electromagnetism, but not Newtonian physics).

And yet, as classical physics crumbled under the revolutions of the 20th century, one relic remained stoutly throughout. The British astrophysicist Arthur Eddington summed up the situation in 1915: 'If your theory is found to be against the second law of thermodynamics I can give you no hope; there is nothing for it but to collapse in deepest humiliation.' Thermodynamics quite remarkably survived both quantum physics and relativity. It now seems that it also survives their unification.

Why, since their origins in the early 19th century, have the laws of thermodynamics proved so formidably robust? The answer lies in the deep connections we have discovered in the 20th century between thermodynamics and information theory — connections that allow us to trace intimate links between thermodynamics and both quantum theory as well as relativity. Ultimately, those same connections show us how thermodynamics in the 21st century can guide us towards a theory that will supersede them both.

Thermodynamics has done astoundingly well for a theory devised purely to explain how steam engines work. The first law states that when energy disappears in one place, it must be found in another. The French engineer Sadi Carnot formulated the second law in 1824 to constrain how much heat an engine could be expected to convert into useful work. Empirically, some waste heat always flows into the cooler surroundings, so a perfectly efficient engine is not possible. The second law states that there is always some degradation of total energy — some non-useful heat is produced. A few decades later, the German physicist Rudolph Clausius explained this restriction in terms of a quantity characterising disorder that he called entropy. The universe works on the back of processes that increase this disorder, for example, dissipating heat from hot places where it is concentrated to cool areas where it is not.

This reading of the second law predicts a grim fate for the universe: once all heat is maximally dissipated, nothing useful can happen in it anymore, and it will die a 'heat death'. Clausius's formulation also raises the perplexing question of how and why the universe started in a state of comparatively low entropy.

Perhaps, because of such undesired consequences, the second law lived for a long time with a question mark against its legitimacy. It was placed there with striking clarity by the British physicist James Clerk Maxwell (Cambridge, rather than Oxford) in 1867. Maxwell was satisfied that inanimate matter did indeed conform to the second law. In a system isolated from its environment, disorder as embodied by entropy did always increase. Heat always passes from the hotter to the cooler, never the other way round; a neat arrangement of dye molecules in a clump readily dissolves in water and disperses randomly, but those molecules never come back together again.

Maxwell's problem was with life. Living things have 'intentionality': they deliberately do things to things around them in order to make life easier for themselves. Conceivably, they might intentionally try to reduce the entropy of their surroundings and thereby violate the second law.

To physicists, such a possibility is highly disturbing. Either something is a universal law, or it is a sham, or a cover for something deeper (which itself could be a new universal law). But it was only in the late 1970s that the US physicist Charles Bennett, building on work by his colleague Rolf Landauer, and using the theory of information developed by Claude Shannon a few decades earlier, finally laid Maxwell's entropy-fiddling 'demon' to rest. An intelligent being can certainly rearrange things around itself to lower the entropy of its environment. But to do this, it must first fill up its memory, gaining information as to how things are arranged in the first place.

This acquired information must be encoded somewhere, presumably in the demon's memory. When this memory is finally full, it must be reset. Dumping all this stored, ordered information back into the environment increases entropy — and this entropy increase, Bennett showed, will ultimately always be at least as large as the entropy reduction the demon originally achieved. The status of the second law was assured, albeit anchored in a mantra of Landauer's that would have been unintelligible to its 19th century progenitors: that 'information is physical'.

But how does this explain how thermodynamics survived the quantum revolution unscathed? Classical objects behave very differently to quantum ones, so the same is presumably true of classical and quantum information; after all, quantum computers are notoriously more powerful than classical ones.

The reason is subtle, and it lies in a connection between entropy and probability contained in perhaps the most profound and beautiful formula in all of science. Engraved on the tomb of Ludwig Boltzmann, who we met at the start of this chapter, in Vienna's central cemetery, it reads simply $S = k \log W$. Here S is entropy, the macroscopic entropy of a gas for example. On the right-hand side, k is a constant of nature that today bears Boltzmann's name, and log W is the mathematical logarithm of a microscopic, probabilistic quantity W — in a gas, the number of ways the positions and velocities of its many individual atoms can be arranged.

On a philosophical level, Boltzmann's formula embodies the spirit of reductionism: the idea that we can, at least in principle, reduce all our outward knowledge of a system's activities to basic, microscopic physical laws. On a practical, physical level, it tells us that all we need to understand disorder and its increase are probabilities. The equation asks no further questions about the nature of the underlying laws; we need not care if the probabilities are classical or quantum in origin.

There is an important additional point to be made here. Probabilities are fundamentally different things in classical and quantum physics. In classical physics, they are 'subjective' quantities that constantly change as our state of knowledge changes. The probability that a coin toss will result in heads or tails, for instance, jumps from ½ to 1 when we observe the outcome. If there were a being who knew all the positions and momenta of all the particles in the universe — known as a 'Laplace demon' (another demon — we like our demons), after the French mathematician Pierre-Simon Laplace who first countenanced the possibility — it would be able to determine the course of all subsequent events in a classical universe, and would have no need for probabilities. This is because the laws of classical physics (the only ones known to Laplace) are completely deterministic.

In quantum physics, however, probabilities arise from a genuine uncertainty about how the world works. States of physical systems in quantum theory are described in what Schrödinger called catalogues of information, but they are catalogues in which adding information on one page blurs or scrubs it on another: knowing the position of a particle more precisely means knowing less well how it is moving, for example. Quantum probabilities are 'objective', in the sense that they cannot be entirely removed by gaining more information.

That casts thermodynamics — as originally, classically formulated — in an intriguing light. There, it is little more than ignorance written down in the form of an equation. It has no deep physical origin itself, but is an empirical bolt-on to express the otherwise unaccountable fact that we cannot know or predict everything that is going to happen, as classical dynamical laws suggest we can. In a quantum context, however, probability and uncertainty are seemingly hardwired into the fabric of reality.

Thermodynamics, rooted in probabilities, acquires a new, more fundamental physical anchor. Together with Müller and Dahlsten, I have looked at what happens to thermodynamical relations in a generalised class of probabilistic theories that embrace quantum theory and much more besides. There too, the crucial relationship between information and disorder, as quantified by entropy, survives.

As for gravity — the only one of nature's four fundamental forces not covered by quantum theory — it might prove to be little more than thermodynamics in disguise. Take all these hints together, and we can begin to see what makes thermodynamics so successful. The principles of thermodynamics are basically information-theoretic, and information theory tells us nothing more than how we interact with the universe and construct our theories. Therefore, thermodynamics just has to be right. It is simply an embodiment of the way we interact with the universe and develop our understanding of it. It is, in Albert Einstein's term, a 'meta-theory': one constructed from principles over and above the structure of any dynamical laws we devise to describe reality's workings. In that sense, we can argue that it is more fundamental than either quantum physics or general relativity.

If we can accept this and, like Eddington, put all our trust in the laws of thermodynamics, I believe it may even afford us a glimpse beyond physics' current order. It seems unlikely that quantum physics and relativity represent the last revolutions in physics: the arrival of new evidence could at any time foment their overthrow in turn. Thermodynamics could help us to discern what such a usurping theory must look like.

A couple of years back, two of my colleagues in Singapore, Esther Hänggi and Stephanie Wehner, showed that a violation of the quantum uncertainty principle — that idea that you can never fully get rid of probabilities in a quantum measurement context — implies a violation of the second law of thermodynamics. Beating the uncertainty limit means extracting extra information about the system (by measuring it), which requires the system to do more work than thermodynamics allows a system in the relevant state of disorder to do. So, if thermodynamics is any guide, whatever any post-quantum world might look like, we are stuck with uncertainty.

After my amble through Oxford, I go out for a drink with physicist David Deutsch, also a colleague of mine. David told me he thinks we should take things much further (in a science sense, not personally). Not only should any future physics conform to thermodynamics, but the whole of physics should be constructed in its image. The idea is to generalise the logic of the second law as it was stringently formulated by the German-Greek mathematician Constantino Caratheodory in 1909: that in the vicinity of any state of a physical system, there are other states that cannot physically be reached if we forbid any exchange of heat with the environment.

I can explain this a bit better using the pint of beer in front of me (David does not drink). In the 19th century, James Joule, a Mancunian brewer whose name lives on in the standard unit of energy, studied the properties of thermally-isolated beer. The beer was sealed in a tub containing a paddle-wheel that was connected to weights falling under gravity outside the tub. The rotation of the paddle warmed the beer, increasing the disorder of its molecules and therefore, its entropy. But however hard we might try, we simply cannot use Joule's set-up to decrease the beer's temperature, even by a fraction of a millikelvin. Cooler beer is, in this instance, a state beyond the reach of physics (except in the pub).

The question is whether we can express the whole of physics simply by enumerating which processes are possible and which are not. This is very different from how physics is usually phrased, in both the classical and quantum regimes, in terms of states of systems and equations that describe how those states change in time. The blind alleys down which this approach can lead are easiest to understand in classical physics, where the dynamical equations we derive allow a whole host of processes that patently do not occur — the ones we have to conjure up the laws of thermodynamics expressly to forbid. But this is even more so in quantum physics where modifications of the basic laws lead to even greater contradictions.

Reversing the logic allows us again to let our observations of the natural world take the lead in deriving our theories. We observe the prohibitions that nature puts in place — be it decreasing entropy, getting energy from nothing, travelling faster than light or whatever. The ultimately 'correct' theory of physics — the logically tightest — would be the one from which the smallest deviation would give us something that breaks those taboos.

Will that be enough to guide us towards an ultimate theory of everything? That is the exciting open question. There is in fact an enthusiastic group of physicists, such as the Brit Lucien Hardy, the American Chris Fuchs and the Italian Chiara Marletto, to mention just a few, who believe that quantum physics could be fully derived from some deeper information-theoretic principles.

Meanwhile, other advantages of recasting the language of physics in thermodynamic terms of possible and impossible transformations are apparent. Not least is that it dispenses with that question of why our universe started as it did, in a state of low entropy. If physics is not at its heart about states and how they evolve, but about what is and is not allowed, that ceases to be a question. In fact, and related to this, the whole notion of time, which currently features prominently in physics, would become of secondary importance.

The barman calls 'time'. According to Deutsch and his collaborator Marletto, time would only enter through transformation involving objects called clocks (anything with a reasonably periodic motion) and there would be nothing more fundamental to it (although I do not think that will go down too well when trying to sneak in an extra pint). Such an approach would probably please Einstein, who once said, 'What really interests me is whether God had any choice in the creation of the world'. While it might not directly answer that question, if a thermodynamics-inspired formulation of physics bears fruit, we are faced with a new certainty: God would simply have no choice but to be a thermodynamicist. That would be a singular accolade for those 19th century masters of steam: that they stumbled upon the essence of the universe, entirely by accident. The triumph of thermodynamics would be a revolution by stealth, 200 years in the making.

So, what might this mean for physics? I look up to the stars on my way home. They are unusually clear, given that I am in the city centre. The moon is not visible either, making the constellations appear even closer. Well, we might just be able to explain how the universe began. And if we can explain that, we can explain the chemistry that made things happen, and the biology that explains life. The Great Reduction would, by definition, bring this all much closer to home.

The current favourite theory for the very early universe is inflation, according to which, the universe underwent a stupendously quick expansion round about the time of 10 to minus 34 seconds. Its diameter grew by a factor of 10 to the power of 30. What evidence do we have that this happened? None at this moment, but the idea helps solve three big — and at first sight — unrelated, observational problems.

First of all, the standard big bang, without inflation, predicts that we should observe magnetic monopoles. These are the magnetic equivalents of electrical charges (which exist), but the problem is that no one has ever observed one. This is known as the 'no magnetic monopoles problem'. The problem is there because the theory is conflicted by the observation.

The second problem it resolves is the 'flatness problem'. On large scales, our universe seems to be remarkably flat. The flatness of this degree would need the initial expansion of the universe to be so fine-tuned that it seems incredibly unlikely.

The final problem is known as the 'horizon problem'. Distant parts of the universe are actually at the same temperature, the so-called Cosmic Microwave Background of about 2.7 Kelvin. But these parts are so far from each other that no signal could conceivably have travelled between them since the birth of the universe. The question then is how different parts of the universe managed to end up equally hot if they could never exchange energy with one another.

Inflation is a clever act of desperation, because it solves all three problems mentioned above at one stroke, despite the lack of mechanism to explain it. The monopoles problem is solved because when the universe stretches by such a huge amount, the density of monopoles becomes so small that they effectively become unobservable. Stretching also makes the universe flatter. Much like the earth, being a big ball, looks locally flat to us (you can see why people held the view that the earth is flat for so long), so does the universe when it has expanded significantly. Finally, the inflation says that the parts of the universe that presently cannot affect one another were in the early times much closer to each other. This explains why the universe looks remarkably uniform.

However, despite these three successes, we actually have no clue how inflation could happen. What caused it and why did it start at such an early stage of the universe? This tells us that we need more astronomical observations on the early universe, but it also probably tells us that there might be a new theory lurking beneath inflation. And perhaps... the Great Reduction that merges micro (quantum) physics with macro (gravitational) physics will also explain the inflation. It all starts with an act of desperation.

Our physics gap, speculations aside, is between atomic behaviour and that of galaxies — but, there is, of course, a great deal of stuff going on in-between. This is the stuff of chemistry and biology, each of which introduces new gaps in our understanding of different natures and potentially different origins. Let us take a flight across our round planet, to China.

CHAPTER II

CHEMISTRY AND COMPUTING: LOST IN TRANSLATION

The chemist at Will's dinner party with the acid green eyes, told us that chemistry's biggest problem is translating what they know about molecules in drug-making into the bigger picture of effects on living organisms. The biggest difficulty? That her computer cannot cope with the micro to macro.

When Richard Feynman was asked what single message he would choose to communicate to any possible life elsewhere if the whole earth was to blow up, he said: that everything is made up of atoms. That would certainly give the recipients of his message a head start (if they understood English, of course).

Sitting outside Beijing's Blue Frog restaurant, a good place to people-watch in the buzzing, drinking heart of the capital, I am humbled again into insignificance, as I am every time I think of this (which is a lot). The laptop I am typing on right now is simply a collection of atoms. The beautifully-crafted heavy oak table underneath it — just atoms. Even the cigar I am puffing on is nothing more than this (even if it does have something of the divine in it). And yet of course they all appear vastly different. In fact, there are only a relatively small number of different atoms, and everything is a combination of them. We are talking a hundred or so different species of atoms. That is it. It is scary when you can reduce everything around you to combinations of a hundred or so *atoms*.

The logic behind this was first uncovered by Mendeleev, a nineteenth-century Russian chemist. He constructed what is known as a periodic table of elements. The idea for a periodic table apparently came to him in a dream. 'I saw in a dream a table where all elements fell into place as required. Awakening, I immediately wrote it down on a piece of paper. Only in one place did a correction later seem necessary.'

The periodic table reflects the fact that atoms that chemically behave similarly actually lie in the same column. This makes sense, but what does it mean when we say that atoms behave similarly? Well, loosely speaking, there are atoms that easily combine with other atoms which we would therefore call chemically active. There are those that are at the other end of the spectrum, being chemically inert. It is impossible to get them to combine with other atoms. It may be the difference between my teenage son and me as a teenager. He is active and sociable and is rarely away from his friends (or his iPhone — his portal to them). I, on the other hand, had my head buried in books and found it impossible to "combine" with other humans. This is how Mendeleev classified them to come up with the periodic table.

But how can atoms (or teenagers) be so chemically different? The reason — unknown to Mendeleev, who did this 50 odd years before quantum physics — is the structure of atoms. Every atom is composed of an electrically-positive nucleus and a varying number of electrons orbiting around the nucleus. These electrons exist in well-defined orbits and each orbit can take at most two electrons. The reason for two electrons is the Pauli Exclusion Principle, telling us that no more than one electron can occupy a single state. When they are on the same orbit, one is spinning clockwise and the other counter-clockwise. The atoms interact chemically by exchanging electrons. If the outermost orbit of one atoms is full, namely if it contains two electrons, then this atom cannot react since it is incapable of receiving more electrons.

Mendeleev's real genius, and this is true for all science theories, is that he was able to use his logic to deduce the existence of six more atoms which were undiscovered at his time. This is the true spirit of science: not only does a new theory need to explain all the current observations, but it also needs to be able to make predictions which were impossible within the previous theory. Unfortunately, this latter predictive element becomes increasingly difficult as a phenomenon gets more complex.

The first reduction between physics and chemistry was, we could say, written in the stars. Modern physics of cosmology also tells us about how different elements were created just after the Big Bang. This is called the Big Bang nucleosynthesis and it was meant to have occurred about three minutes after the Big Bang (hence, the title of Weinberg's famous popular book on this topic, *The First Three Minutes*). At that point, we had protons and neutrons, and the temperature was stupendously high. Their interaction lead to the light elements such as hydrogen, deuterium, helium, lithium and perhaps boron (I say 'perhaps' because we are not sure). This all ended about 20 minutes after the Big Bang and the temperature dropped so much that we know that heavier elements were exceedingly unlikely to have been created. This is because to get larger atoms, the nuclei of smaller ones need to merge together, which requires overcoming the very strong electrostatic repulsion between the positively-charged protons.

In fact, the heavier elements were and are all still being created in the stars, which themselves were formed by the gravitational collapse of hydrogen and helium. This is why Carl Sagan used to repeat beautifully that we are all made of stardust — when the first generation of stars exploded, all heavy elements generated therein were subsequently blown off into the universe. The iron in our body, for instance, was created in these early stars. Sagan also talked of the spiritual aspect of this discovery: we are literally made of the same stuff: all is one and one is all.

And yet, physics nearly caused very big problems for chemistry, for it is when we first looked at atoms quantumly that the biggest gap between physics and chemistry was found. The bottom line is: atoms, and therefore molecules, could not exist in the classical world. That would mean no laptop, no cigar. No book. The reason why classical laws cannot explain the stability of atoms is that classical physics predicts that as charges accelerate, they have to radiate energy. So, since an electron (which is a negative charge) moving around a nucleus inside an atom is in a state of constant acceleration around the nucleus, classical physics expects it to be losing energy by constantly radiating it away. As it continues to do so, the electron would have to move closer and closer to the nucleus (to which it is electrically attracted since the nucleus is positively charged) and ultimately fall into it. So, of course, no atom and hence, no molecule could exist in the classical world.

Fortunately, quantumly, atoms are 'saved' from the (classical) collapse by what is known as the Heisenberg uncertainty principle. This principle is the corner-stone of quantum physics and it tells us that we cannot know the position and the speed of any object perfectly accurately. The more we know *where* something is, the less we know about how quickly it is moving, which implies a larger energy. So, if we know exactly where the electron is, we would then not know its speed and therefore its energy (the electron would be in a superposition of many different speeds). Intuitively speaking, this energy is what keeps the electron from collapsing into the nucleus (as collapsing into the nucleus is tantamount to knowing the electron's position really well).

This is the same Heisenberg whose 'cuts' I had hoped to speak of in the Oxford dinner party, before everyone started to agree with me. This 'cut' is meant to be the dividing line between the classical and quantum world. For Heisenberg, the cut was not a hard boundary, but more of a fluid concept that could be moved to encompass larger and larger objects. The main gap we are exploring here, between the physics and chemistry, could be thought of as a Heisenberg cut (and perhaps so could any other micro to macro chasm). I cannot help but add in a story of Heisenberg and his cuts, especially when we talk of larger objects.

Heisenberg has been immortalised in the public image by Michael Freyn's play *Copenhagen*. This play is a fictitious dialogue (but which is likely based on some truth) between Heisenberg and his mentor Niels Bohr (a Dane, native of Copenhagen and about to flee to America ahead of WWII) regarding the Nazi effort in making an atomic bomb. Unlike many a German scientist who escaped from Europe after Hitler's rise (often to Oxford), Heisenberg stayed in Germany, leading the effort to make the bomb. After the war, of course, Heisenberg tried to insinuate that he actually deliberately slowed down this programme, sabotaging it to aid the Allies. On top of it, the claim was that he alerted Bohr to the fact that Germany was engaged in such an activity which prompted Bohr to urge the Americans to do the same (which they did — and beat the

Germans to it). The other side of the argument — and there is always the other side in such complex historical scenarios — is that Heisenberg did no such thing, but (luckily for us) made a simple overestimate as far as the amount of Uranium needed to make the bomb. So, he saved atoms from collapsing (classically). Perhaps, he saved Europe from collapsing too.

I have been thinking quite a lot about all that Heisenberg did for the dividing line between quantum and classical physics, which makes something that happened at a conference last week, just before I came out to Beijing, particularly amusing. So, this conference was about complexity, and one of the speakers — a medical doctor called Stu Kauffman — proposed that the region between quantum and classical physical domain could be real. He did not just propose it, as we found out at the end when the last screen popped up — Kauffman had actually *patented* this region, a*nd any technologies that might be based on it!* I kid you not. It really is possible to own different domains of reality. Perhaps he is more of a genius than he first appeared. I mean, I am all for bravery and even acts of desperation, but for a medical doctor to patent a region that (almost certainly) does not exist is not just a leap, it is jumping out of a plane with no parachute. Everything, as far as we know, is quantum, it is just that under some conditions, classical physics describes the behaviour well enough that we do not need to use quantum physics. Rant over. Poor Heisenberg would be turning in his grave.

By the by, being able to, in Heisenberg's way, explain the chemical periodic table, quantum physics closes the biggest gap between physics and chemistry. Quantum physics can explain atoms as we see them now. But, can quantum physics go further than this and explain how it is that atoms combine? How atoms chemically interact to produce different outcomes and behaviour?

Classical physics finds it impossible to explain how even the simplest of atoms — such as hydrogen — combine to form a molecule. Rather than tell you about all possible types of chemical bonding (for it would take us rather off-track), I will explain how one hydrogen bonds with another hydrogen, when it has lost its only electron. This is the simplest molecule we can think of, but already subtle as far as its physics is concerned.

As these two atoms approach one another, the electron from one of them now has a chance (remember, quantum physics is all about chances for things to happen) to hop to the other one (which has lost its electron and therefore has a vacancy). As the electron hops, it also has a chance to hop back. In fact, quantumly, the electron can be in two different places at the same time, close to one hydrogen atom and close to the other one.

There is a well-known and beautiful little experiment that shows the interchange of energy between two vibrational systems. Two pendulums are hung up together, side by side, on a string that is stretched horizontally between two fixed supports. If one pendulum is set in motion, swinging to and fro, then its motion will gradually die down, while, at the same time, the second pendulum begins to move, and swings with an ever-increasing amplitude, until the first pendulum comes to rest. Then the process will be reversed, the motion of the second pendulum dying away and the first increasing. The process will continue until both pendulums are brought to rest by the resistance of the air and other frictional forces. And in the same way, an electron can hop between two atoms.

It turns out that the state when the electron is in this kind of quantum superposition is energetically lower than if the electron is localised in just one atom. Since nature favours states of lower energy, the two hydrogens end up bound. The binding is therefore mediated by the electron being in a superposition, something that is classically speaking completely impossible. This gap between chemistry and quantum is not just bridged, they are dependent upon each other.

Nice explanation, but why does the same not work for helium? Why do we not ever see two helium atoms form a molecule? The reason is that helium has two electrons in the same state and cannot receive another one (due to Pauli's exclusion). So electrons are prohibited from hopping between two helium atoms and there is, therefore, no way to reduce energy by forming a molecule.

The dynamics of how atoms combine, even simpler ones with just a handful of electrons, is actually so intricate that it is difficult to simulate quantum mechanically in all its nitty-gritty. Here, I have in mind even the binding of two hydrogen atoms into a hydrogen molecule, or two hydrogens and an oxygen combining to give us a water molecule. The full details of these processes are still open to research, even though it is clear quantum mechanically why these reactions occur in nature. We are just waiting for the supporting evidence.

There is an interesting recent twist in the story of the water molecule. As I type this, the waiter in the Blue Frog, which is now becoming a little busier as day fades into night, attentively fills up the small rectangular glass next to me with iced water. He sets down my single malt next to it. Water is one of the most unusual liquids, and many of its quirks are essential to life. For example, its higher density as a liquid than as a solid means ice floats on water, allowing fish to survive under partially frozen rivers and lakes. It also has a high heat capacity, meaning that it takes a lot of heat to warm water up even a little, a quality that allows mammals to regulate their body temperature.

But computer simulations show that quantum mechanics nearly robbed water of these life-giving features. Most of these features arise due to weak hydrogen bonds that hold H_2O molecules together in a networked structure. For example, it is hydrogen bonds that hold ice molecules in a more open structure than in liquid water, leading to a lower density. By contrast, without hydrogen bonds, liquid molecules move freely and take up more space than in rigid solid structures.

Yet, in simulations that include quantum effects — and here we are forced to do computer simulations due to the complexity of the problem — hydrogen bond lengths keep changing (thanks to the Heisenberg uncertainty principle), which says no molecule can have a definite position with respect to the others. This de-stabilises the network, removing many of water's special properties. Surely this leads us to question whether quantum physics can explain the existence of the water molecule? How water continues to exist as a network of hydrogen bonds, in the face of these de-stabilising quantum effects, was a mystery until quite recently.

I take a sip.

In 2009, quantum physicist Thomas Markland suggested a reason why water's fragile structure does not break down completely. Together with his colleagues, he calculated that the uncertainty principle should also affect the bond lengths within each water molecule, and proposed that it does so in such a way as to strengthen the attraction between molecules and maintain the hydrogen-bond network. 'Water fortuitously has two quantum effects which cancel each other out,' Markland says. Yes, there is an uncertainty principle about the location of each molecule, because of the Heisenberg uncertainty principle telling us that unless the speed of the molecules is arbitrarily large, we cannot know their exact location. However, the molecular bond lengths which affect the forces between atoms in the molecule also suffer from the same uncertainty principle.

This turns out to exactly cancel the first atomic position uncertainty. Thank goodness.

There are still, of course, some questions in chemistry that are hard to reduce to quantum physics, and we still have a great deal of explaining to do. But as with every gap that has thus far been revealed between the two natural sciences, it is simply a matter of time before these questions are answered. There are certainly no indications that anywhere within chemistry will the laws of quantum physics actually break down. There are also no indications that the laws of quantum physics will not be sufficient to account for all the chemical phenomena. Ergo, there is no real gap, save for the one in our own limited experiments.

Sadly, this is not the end of the story. It is precisely our limited experimentations and the reasons behind this that are responsible for the biggest gap between the micro and the macro in chemistry: computers. As the chemist with the acid green eyes told us.

COMPUTERS: CHEMISTRY'S BIG GAP

Quantum calculations in chemistry are almost always — due to the complexity of the situation — done numerically on a computer. Here, therefore, we encounter another intriguing possibility. Could it be that the gap between physics and chemistry is inherently irreducible just because we might not have enough computational power? Unlike when we talked about quantum physics and gravity, here the irreducibility would be of an entirely different origin, computational in nature.

All this, interestingly enough, is related to the following question.

A barber in a small village shaves all men who do not shave themselves. Question: does the barber shave himself?

I will give you a bit of time to think about this conundrum if you have not seen it before.

Now, as you have hopefully realised by this point, the book you are reading is about sciences and the gaps in our understanding of nature that each of them invariably contains. Computing, on the other hand, deviates slightly from our remit, in its grounding in mathematics. One of the first proper scientists, Galileo Galilei, said of the relationship between mathematics and sciences: 'The great book of nature is written in mathematical symbols.' In other words, we need mathematics in order to explain natural phenomena. It helps us, of course — but could it also hinder us? Be the reason behind our gaps and hence, the difficulty in the Great Reduction?

Mathematics is compartmentalised in a way similar to the sciences themselves. This starts at school, where we learn, for example algebra, arithmetic and trigonometry as if they are not interconnected. It saddens me how this separation turns a lot of school children off mathematics, as they are often not shown the relevance of maths to everyday life. And wider than schooling: it could easily be that an expert in number theory, for instance, knows very little about functional analysis.

Ah, would it not be pleasing if all of mathematics was reduced to just one set of rules? This would close all the gaps between different types of mathematics. After all, this is what we are asking about sciences in this book, so why not ask the same of mathematics? Unfortunately, however, the idea to reduce all mathematics to simple axioms was blown to pieces by an Austrian logician called Kurt Gödel. Thanks, Gödel.

Luckily, it is hard to stay angry with Gödel. With a personality straight out of a Kafka novel, and naivety about the real world that only a real abstract thinker can display, he left behind a legacy of anecdotes. One story says that when he was crossing the US border while running away from the Nazis, he challenged the immigration officer regarding the American constitution. The officer said that unlike the Austrian and German constitutions, the American one prohibited dictatorship. Gödel then said: 'That is not true. I have read the American constitution and it too allows dictatorship. Here is how that can happen…' Luckily, before he could detail his argument, his friend (another European émigré) pulled him down and told him to shut up. But, endearing as he may have been, he did put a spanner in the works as far as mathematical (and therefore, computational) (and therefore, scientific) unification is concerned.

Before we get to Gödel's discovery, we must journey back before his time, to when the story of trying to unify mathematics begins. Enter British mathematician (and later philosopher) Bertrand Russell. Towards the end of the nineteenth century, Russell was working with another mathematician, Alfred Norbert Whitehead, on reducing all mathematics to logic. This is underpinned by Georg Cantor and Richard Dedekind's set theory a few decades before.

As the name suggests, set theory is about sets — which are collections of objects. A set of natural numbers contains all the numbers 1, 2, 3 and so on. Every number could be thought of as a set too. Number 3 could be the set of all sets containing exactly three things (three sheep, three cigars — yes, please — or the Father, the Son and the Holy Ghost). In fact, everything Russell and Whitehead encountered could be seen as a set, and the whole of mathematics — be it trigonometry, algebra, analysis — could be reduced to relationships between these sets. What a wonderful vision! The whole of mathematics reduced to set theory!

Alas, it did not last.

Russell and Whitehead worked hard on phrasing the rest of mathematics using sets. Just as they were about to reach completion — some 10 years after they had begun — Russell came across a horrifying realisation. He came up with an example of something that could not be thought of as a set.

Imagine a set of all sets. This set has to contain itself, and this is simply true by definition since it contains *all* possible sets. Now imagine a set of all sets other than itself. This set differs from the first only in one element: it does not contain itself though it does contain all the other sets.

So there are sets that contain themselves and those that do not. For instance, the set of all physicists is not itself a physicist and therefore it is a set that does not contain itself. On the other hand, a set of all imaginary things is a member of itself since it is itself imaginary (by definition!).

Finally, imagine a set of all sets that do not contain themselves. These are sets of physicists, sets of plants and so on. Question: does that set contain itself?

Back to our barber and his potentially hairy chin. If you realised what the problem was with the barber shaving himself (or not), the same applies here. If we suppose that the barber shaves himself then he ought not to, since he only shaves those people who do not shave themselves. But if he does not shave himself then he should, since he shaves all those who do not shave themselves. There is no way out of this but to infinitely flip between the two possibilities. Neither can be right.

The same goes for the set of all sets. If we suppose that the 'set of all sets that do not contain themselves' actually contains itself, then it should not since it only contains those that do not contain themselves. But if it does not contain itself, then it ought to, since it contains all those that do not contain themselves!

This led poor Russell into depression, since he had found a class of things that cannot be treated as sets, and therefore discovered an internal inconsistency with reducing the whole of mathematics (or anything else for that matter) to the rules of logic. In an act which is the mathematical equivalent of a religious conversion, Russell became a philosopher. Disclaimer: I am certainly not suggesting that becoming a philosopher is a good way to deal with depression (though it seems to have worked for Russell).

So, Russell's discovery was the first glimpse that mathematics might not be reducible to one of its branches. What Gödel realised, a couple of decades later, was that *none* of the branches of mathematics can ever be reduced to their respective simple axioms (another name for a mathematical self-evident truth). Gödel, in a nutshell, mapped mathematical truths into numbers. For example, I have a glass of single malt in front of me. If I were to ask for another glass (not a bad idea), I would have two glasses, or double the single malt I had before (and perhaps double the hangover). So, $1 + 1 = 2$. This single derived mathematical truth (a.k.a. a theorem), Gödel discovered, could not necessarily be derived. So not only are different branches of mathematics, like arithmetic, not reducible to logic, but even logic or arithmetic cannot be fully explained using simple axioms.

Rather than going any deeper into Gödel's argument, I will talk about an equivalent argument using computers — which are, for most of us, much easier to grasp. Let me call to the stage a certain behind-the-scenes war hero, Alan Turing. Spoiler alert: the big difficulty that Turing ran into will have far-reaching consequences for the gap between physics and chemistry, as well as physics and other sciences.

You might wonder if Russell's and the barber's paradoxes are just mind games. It seems like there is trouble if you make definitions self-referential, but does this actually matter in practice? Well, it certainly does, and Turing is the person to credit for this realisation. (Turing is the person to credit for quite a lot, actually; his code-breaking work was estimated to have shortened the war in Europe by as much as four years.)

Not long before his work in World War II, Turing invented the idea of universal computers in a laboratory in Cambridge. They are computers that can, in principle, simulate all other computers. All our current computers are really universal, so this idea is not strange to us. But in Turing's time — before computers actually existed — this was a revolutionary idea. Turing was a visionary and he introduced this concept because he believed that these kinds of computers could actually simulate human thinking.

However, just like Russell's sets, universal computers had a limitation. They could not decide if a randomly imputed program would ever stop. This paradox is known as the Halting problem, and is similar in nature to Russell's. There are programs that when imputed into a computer, such as 'print "I love you" 5 times', will stop, and those that will go into an infinite loop, such as 'tell me if the barber shaves himself?'. But these are obvious and it is easy to tell how they will behave simply by inspection. However, most programs are much more complicated and it turns out that the question of whether they will ever stop or not is ultimately undecidable.

In the same way that Russell's sets can capture most things except sets of sets, universal computers can simulate any other computers and their behaviour, but are still incapable of solving the Halting problem. Do not get me wrong. You can always go ahead and run the program to see if it halts. But, if after ten days of running it is still going, you will have no idea if it will ever halt in the future and produce a result. Most programs are in fact so complicated that the only way to see what they will do is to run them and wait.

The intuition as to why computers suffer from a Halting problem is similar to Russell's paradox. We need a program for a universal computer — let's call it the Halting program — that will decide if any of the possible programs that can be fed into the computer will halt. But among any of these programs is also the Halting program itself. But this means that the Halting program has to contain more information than itself! In other words, the Halting program can only address programs that do not address themselves.

I am reminded of a fascinating view of time I once read. According to this theory, there are infinitely many times rather than just the one according to which we all set our watches. I first read of this idea in a book by English engineer and pilot John Dunne — 'An Experiment with Time'. The crux of his argument is that we need to treat time as another dimension, and we can move up and down the time axis just like we can move back and forth in space. However, in order to describe the movement along the time axis, we need a second time axis telling us how quickly we are moving along the first one. But as soon as we do that, we realise that we need another time axis to describe the motion along the second time axis. And so on, ad infinitum.

There is a huge painting covering one wall of the Blue Frog café, where I am still sitting, a few whiskies later. I can see the painting from where I am sat at my usual table outside and I often find my eye drawn to it, although I do not particularly like it as a piece of art. It is a supernova explosion: apt, but something about the proportions does not quite sit right for me. There is another analogy of what we have been talking about — a painter who wants to paint an accurate picture of the world. But as soon as he thinks he has done it, he realises that he himself — the painter — is missing. And no matter how many times he adds an extra painter, the ultimate painter will always be missing from the picture. And so, it is with the Halting problem — we can never have a program that decides if a random program will halt.

The way that Turing explained the Halting problem was to use a method that Georg Cantor, the German mathematician whom I briefly mentioned earlier when talking about set theory, invented to show that the number of real numbers is uncountable. In other words, Cantor proved that there are more real numbers than integers.

Here is how he did it — a truly beautiful and simple piece of reasoning. A deduction worthy of Mr Sherlock Holmes. Imagine that we can enumerate all real numbers. That means that there is a list, albeit infinite, which might look something like:

1 0 0 0 0 0 0 0 1...
2 **2** 2 2 3 3 3 3 3...
4 3 **5** 6 7 8 9 2 0...
7 6 8 **8** 3 9 3 9 9...
6 7 4 8 **3** 9 3 0 2...
5 7 3 9 9 **2** 7 4 9...

And so on. Now, Cantor says, take the first digit of the first number, the second of the second number and so on. This gives us the number **125832**...now add 1 to every digit to obtain **236943**...(adding 1 to 9 gives 0 by agreement). The last number cannot be on the list since it differs from the first number on the list in the first digit, it differs from the second number in the second digit and so on. Therefore, we have found a number that is not on the list, which conflicts with the assumption we started with that the list enumerates all real numbers.

Whatever list we enumerate, we can always use Cantor's diagonal argument to construct a number that is not on that list! Thinking of these numbers as numbering different computer programs leads us to the Halting problem.

Sometimes thoughts come unexpectedly while we are engaged in a different activity altogether. They say a cure for writer's block is to go and do something entirely different (although I am a little too comfortable right now). When the mind and body is engaged elsewhere, deeper thoughts can arise from a place of no pressure. Showers — many people say that breakthrough insights came to them during a lovely blast under high-pressured water. Shinto monks engage in purification rituals and meditations under waterfalls, known as Misogi. The shower has the same effect as waterfalls (although they are slightly colder) that stimulate your thinking.

For me, this kind of insight came at the Great Wall of China. I visited this incredible structure yesterday, which stretches over 7000 kilometres of mountain ranges in northern China. Built during the Qin dynasty (221–206 BC) — the one which unified China — with the intention of providing defence against northern invaders. The view from the wall is breathtaking and I cannot help but marvel at the amount of time, effort and human life that was consumed during this project.

During the five-hour-long hike, I had a most interesting discussion about the ancient Chinese philosophy of Taoism with a colleague from Beijing who also happens, I discovered on this walk, to be Taoist. Tao, roughly translated, means 'the way', and this philosophy can be described as 'go with the flow': do not try, just be. The phrase that captures this is wu-wei, something that in our own words could be thought of as 'being effortlessly calm'.

There is a deep philosophy behind the idea that we should lead our lives in this way. The most famous Taoist text, *Tao Te Ching*, written by Laozi sometime in 6th century BC, opens with the following lines:

The Tao that can be told is not the eternal Tao.

The Name that can be named is not the eternal name.

It occurred to me during the Great Wall hike, that perhaps this is the Taoist way of phrasing the Halting problem, and the limitations of any language, including Gödel's incomplete formal mathematical system. When you name something, you have automatically limited it. As soon as you limit it, you can think of something else that lies beyond this limit. A very young child marvels, open-mouthed, at the beauty and depth of nature. But they soon learn to label it: 'flower', 'tree'. After this labelling, what they see slowly loses its magic as the world is reduced to signposts. Perhaps, we lose the concept of infinity through our desire to classify and compartmentalise.

Halting and Gödel's and Cantor's arguments crucially rely on the concept of infinity, namely the fact that we can always add more numbers. In a finite system (for instance, any PC) we will run out of space sooner or later. If the memory is finite, we will only be able to list a finite number of things before we run out of space. Then Cantor's argument no longer applies. In Turing's equivalent, the integers are like the programs which decidedly halt or not. Given them, we can always construct a program that is undecidable by an argument that mirrors the Cantor diagonal argument.

Do programmers worry about this? 'Oh, you bet,' one of my oldest university friends told me, a programmer himself, rolling his eyes. One way of dealing with Halting, he explained, is to make sure that every subroutine in your program code is actually executed before a certain pre-prescribed time. Another is to use restricted programming languages that are known not to lead to infinite loops. Here the Halting problem has a major impact in the real world. If you consider writing computer programs a real world, that is. 'Perhaps some would say this is becoming the only real world', he suggested, provocatively. I would not tell you my reply.

Now, the punchline, as far as the reduction of chemistry to physics is concerned. Summoning back our chemist lady from the Oxford dinner party with the acid green eyes, I recall her despair as she told me that most of the problems in chemistry, such as 'compute what happens when such and such molecules collide', are too intricate to be solved by hand. They have to be simulated numerically on a computer.

Suppose I give you some ingredients to a chemical reaction and ask you to tell me if the concentration of nitrogen will be bigger than the concentration of oxygen at any point in the future. This problem is the same as the Halting problem and cannot be answered — although we know that the answer has got to be either a definitive 'yes' or a definitive 'no', there is no way of telling other than to watch the reaction unfold.

This was actually formally proven by a colleague of mine, Mile Gu, in his PhD thesis. He showed that if you simulate chemical reactions on a computer, then solving the question of different concentrations ends up being the same as solving the Halting problem.

As it happens, I am writing these lines in a café near the famous Tsinghua University, where I will visit Mile this very afternoon. He is a professor of quantum information here (though soon to move to Singapore), and his research is an important piece of the puzzle.

And so to the biggest question: can quantum physics explain, at least in part, the Halting problem? Because if it *can*, and this big computational conundrum can be resolved, then there would no longer be the gap between physics and chemistry. And that would solve a heck of a lot of our problems.

The area around the university is full of life — noisy, smoky, dirty and loud. I love it. I am looking outside of the window of the Bunny Drop café where I am sipping at my double espresso and drinking a carrot juice on the side (when in China…).

Close by are the beautiful Summer Palace Gardens and the magical Yuanming Yuan park, housing the Old Summer Palace. Walking around the waterlily-sprinkled lakes is an almost spiritual experience. Crossing small stone bridges to discover different alleys and quads makes you experience what Chinese emperors throughout centuries were enjoying alone. There is also, disconcertingly, a reminder of some of the savagery and looting during the Opium Wars of 1839–1860 and subsequently, by the Cultural Revolution.

The crucial ingredient in Mile's thesis was the so-called Ising model. It is a very simple model, but Lars Onsager received a Nobel Prize for solving it in two dimensions. Already, that was considered exceedingly difficult.

Suppose that we have a chain of systems, each of which only has two states; 'up' and 'down', and where each system only interacts with its two neighbours. This interaction can cause spins to flip between the two states, from up to down and vice versa.

Such a system was first considered in physics by Ernest Ising, who studied it during his PhD in the 1920s. What interested him was if there was a situation when a chain like that can become spontaneously magnetised, namely whether all spins can suddenly point in the same direction. Unfortunately, he was able to prove very conclusively that there is no phase transition in such a model. So, if you think of atoms in such a chain as being little magnets, then they can never ever spontaneously align with each other, no matter what happens outside. What Ising was doing was also bridging the gap between micro and macro. From atoms as tiny magnets to the whole solid that they comprise being magnetic. The interesting thing is that atoms can spontaneously become magnetised even though, initially, they are all demagnetised. However, this can never happen in a one-dimensional world. And this was apparently such a disappointing conclusion to Ising, who was hoping to explain phase transitions microscopically, that he quit physics after his PhD!

Too bad for Ising, because some 20 years later Lars Onsager showed in a very beautiful paper that if instead of a chain, you look at a two-dimensional array of atoms, then there is a phase transition at low enough temperatures. And for this, Onsager was awarded a Nobel Prize for chemistry. The whole area of phase transitions then started to thrive, resulting in a number of fundamental discoveries and Nobel Prizes.

But for our story, it is important that the Ising system can actually implement universal computation. If you think of states up and down as a zero and a one, and the interaction between spins as a computational gate, then the two-dimensional Ising model is actually a universal Turing machine. It can perform any computation we like, but, it therefore also suffers from the Halting Problem. In this case, the problem is this: given a random initial configuration of systems (some of which are up and some down) can we tell if sometime in the future there will be more systems in the state down than up? The answer is no, we cannot tell, for otherwise we could solve an instance of the Halting problem.

But, is more really different? This is a deep question and it usually goes under the name of reductionism. There are many ways in which we can think of reducing one explanation to another one. For instance, we could say that chemistry is reduced to physics when all the truths in chemistry (this includes all the laws known in chemistry) can actually be phrased using the language of physics. We could, on the other hand, suggest that 'all the observations explained by one theory, can also be explained by another one'.

Neither of these is probably as strong as what we have in mind in physics when we talked about explaining chemistry using quantum physics. What we mean is that all the laws of chemistry can actually be formally — mathematically, I mean — derived from the laws of physics. This is meant in the same way as the derivation of Bernoulli's and Boltzmann's where the macroscopic equation of a gas is derived using the macroscopic laws of motion of Newton. Gu and his colleagues gave a beautiful illustration of a physical system that cannot be easily 'reduced', and of the developing symbiosis between theoretical physics and computer science.

To address 'the understandable' in reductionism, British scientist Stephen Wolfram examined the relation between computation and the unfolding of the physical world. He defined as reducible those systems for which there is a computational shortcut that allows their behaviour to be efficiently predicted rather than reproduced step by step. For example, the motion of a simple pendulum is described by a cosine function that can be computed using a rapidly converging mathematical series, rather than simulating each and every pendulum oscillation. Such shortcuts do not usually exist for chaotic systems, for example.

Wolfram made an additional, important point. Many systems are irreducible, but among them, only a few are undecidable: they have properties that cannot be formally calculated as stated in Kurt Gödel's and Alan Turing's theorems. And this is where the notion of 'different' (or complex) systems can be made more precise — those with undecidable global properties despite having well-understood local (microscopic) governing laws.

As a first example of "undecidability", consider a cellular automaton (CA) — a lattice of cells, each of which can take on a finite number of values (states) and evolves over time according to the configuration of a set of neighbouring cells. This is the microscopic transition rule. For the one-dimensional CA known as 'elementary rule 110', two states are allowed ('0' or '1'), and any cell will evolve to 0 if either its state and that of its right-neighbour cell are 0, or if its state and those of both its immediate neighbours are 1 — otherwise, it will evolve to 1. Thus, the local governing law is fully understood.

But, the global dynamics of a CA is a different matter. Each row displays the lattice at a different time step, thus providing a full spatiotemporal record of the dynamics of the system. Cells far apart act in concert to sustain 'particles': structures that move and interact, and in doing so, compute. The result is an intricate and undecidable global dynamics even though all the rules of the game are perfectly well known.

It is not easy to demonstrate that rule 110 can simulate a universal computer. Such proofs often involve the construction of a few logic gates and information channels that allow universal computation to be implemented, and could well be argued to be reductionist. But once these elements have been constructed, the step that shows that a system has undecidable properties involves proof by contradiction rather than constructive proof: a higher level of abstraction. When Gu and colleagues write "the understanding of macroscopic order is likely to require additional insights", they may have in mind procedures, such as proofs by contradiction, that transcend mere reductionism.

Mile Gu and his colleagues, in fact, focused on the Ising model: a lattice of spins that interact with one other and with an external magnetic field. The individual spin states can be 0 or 1 (corresponding to 'up' or 'down' magnetisation), just like those of elementary CA. The main difference is in the dynamical rule: spins tend to align with their neighbours (and with the external field, if one is applied to the system), whereas thermal fluctuations counteract and randomise their state. Therefore, the microscopic transition rules are probabilistic. As we said, Ising models in more than one dimension exhibit phase transitions: at sufficiently low temperatures, the tendency of spins to align overcomes thermal jiggling, and the system becomes and remains ordered. Perhaps not surprisingly, the physics and mathematics immediately around the disorder-to-order phase transition are rich, and have been well-studied.

In their study, Gu and others mapped the dynamics of a certain CA into the lowest-energy (ground) states of Ising models. They grouped spins into blocks that encode the logic operations needed to produce universal computation in the corresponding CA. They then defined the 'prosperity', p, of two-state systems as "the probability that a randomly chosen cell at a random time step is live" (live meaning state 1).

Using the computational properties of the CA, Mile Gu and colleagues were able to show that p is undecidable for infinite, periodic Ising systems. They argued that, as a consequence, many macroscopic properties of an Ising system, including the system's magnetisation and degeneracy (number of independent configurations) at zero temperature, depend on p and hence, are also undecidable. Because Ising models have been used to describe not only magnetic materials, but also neural activity, protein folding and bird flocking, the consequences of Gu and colleagues' results transcend both computer science and physics.

Alas, their results apply only to infinite lattices, and unfortunately, therefore seem of limited use. The finite Turing systems one would encounter in real life are always decidable. But there are hints that finite objects may, after all, have undecidable properties. One hint comes from certain mappings of a solid square onto itself, which have been shown to be undecidable. These procedures slice and rearrange parts of the square in a way that allows computer operations, such as shifts, to be implemented, and they take advantage of real numbers (which require an infinite number of digits) to pack an infinite computer into a finite region. A second hint comes from a new level of computation that is more powerful than a Turing machine, and has been proposed as just the right one to simulate natural physical phenomena. I wonder, as I wander out from the café to meet with Mile, whether his work, along with these two ideas, will lead to a better understanding of the 'computer' in which we live.

This line of reasoning suggests that chemical systems are inherently more complex than physical systems, and that some of their properties cannot be captured, even in principle, by the laws of quantum mechanics. They have emergent properties that only happen beyond a certain level of complexity, when many tiny subsystems interact with one another.

This does not mean that chemical properties cannot be phrased using the language of physics — this is always possible and in fact, this is pretty much what we did above when describing the problem. But this does mean that there might be an inherent gap between theories that cannot be bridged, even in principle.

This gap could not be bridged even with quantum computers, though it could be made smaller because quantum computers could perform some chemical simulations much faster than classical. However, the Halting issue would — as far as we can tell — still remain. Even quantum computers could not compute the incomputable.

Einstein's oft-quoted words, 'God does not play dice with the universe,' allude to his scepticism that quantum theory is the ultimate descriptor of reality. His opinion is common sense to a realist — every object, every state, and every piece of information — should exist in definite state. Yet, quantum mechanics predicts otherwise, that objects could exists in 'quantum superpositions'; a cat could be simultaneously dead and alive.

A hundred years on, experiments continue to confirm quantum physics — yet Einstein's statement continues to haunt us too. Every piece of information we observe, every experiment we perform, can all be described by classical information — we can (and do) write the description off all experiments and their outcomes down on pieces of paper. So why is nature ultimately quantum mechanical?

A clue to this puzzle may emerge from the most unlikely of places, with links to far more practical concerns — the design of the new cities, the understanding of the economy, the tracking of epidemics, the dynamics of human societies. This appears surprising at first, for these topics seem to have little to do with each other, and much less with quantum theory. Yet, when experts in these fields gathered in Tianjin for the World Economic Forum with unified commitment to "improving the state of the world", it emerged that they had much more in common. We will journey into this in much more depth in the second half of this book, when we move from natural science to social science. Although, the jump may be smaller than you think.

CONCLUSION

Beijing is a perfect place to ponder the micro-to-macro gap and whether this gap will ever be bridged. We are talking about a city with a population larger than most European countries — 30 million — crammed into an area smaller than Luxemburg. Montenegro (look it up, it is where Casino Royal is) has a population 60 (sixty!) times smaller and an area about hundred times bigger than that of Beijing.

And I wonder, will the fraction of the female population in Beijing ever be bigger than male in the future? The excess of men is one of the biggest problems facing China nowadays, since it is well documented that a surplus of young men without many prospects is the best recipe for civil unrest. The Chinese government (and other governments really) worry about this.

Forget that one. It sounds complicated. I am looking at a really busy street in front of the Bunny Drop café, with traffic going in all directions, motorbikes, bikes, cars, busses and pedestrians all negotiating the same interlinked busy paths. By simple inspection, I see that the number of pedestrians on the far side is bigger than on the near side. Will the number of pedestrians on the near side ever be bigger in the future?

It is a simple question. But, yes, you have guessed it. What if all these kinds of questions, even the simple sounding ones, are actually undecidable, like the Halting problem, if we try to answer them by simulating them on a computer.

Wherever you look in Bejing and China, you have got micro to macro questions bubbling up in your imagination.

Quantum Simplicity: in using quantum theory to better understand complexity, we may also open a rare windows of insight to quantum theory itself. If we are to build a computer simulation of a process, then every bit of information that we need in order to make optimal future predictions is a bit of information we must store within our computer. The capacity for quantum computers to record less information without sacrificing predictive power points towards simpler simulations — resulting in an ultimately simpler view of our entire reality.

The appeal of simplicity has long been an almost universal human aesthetic. Indeed, in the words of Isaac Newton, 'We are to admit no more causes of natural things than such as are both true and sufficient to explain their appearances,' a statement that has guided scientific development since its inception. In answer to Einstein, could we say that God plays dice because he too shares our appeal for simplicity?

CHAPTER III

BIOLOGY: THE BIGGEST GAP OF NATURAL SCIENCE

One of the most interesting questions I have ever been asked came after a lecture I gave in memory of my late friend, Peter Landsberg. Peter and I met back when I was a PhD student, both giving talks at a conference in Nottingham. Peter's style was inimitable. His sense of humour and quick wit were a breath of fresh air amid some pretty boring talks.

In Peter's memorial lecture, I talked a lot about his very interesting (and provocative) written exchange with another physicist, Eugene Wigner, regarding quantum physics and the origin of life. Wigner had written a controversial paper claiming that quantum physics cannot explain the origin of life. More precisely, he said that it is exceedingly unlikely that initial conditions would allow quantum physics to lead to anything resembling life. This paper is still quoted by many creationists to argue that evolution cannot explain life. Peter Landsberg challenged this in an exchange published in science journal *Nature* and showed Wigner's argument to be wrong.

Now, this was rather a big deal. Wigner was, after all, a Nobel Prize winner. It is not every day that Nobel laureates get things wrong *and* get found out (despite Niels Bohr's opinion that an expert is a man who has made all possible mistakes which can be made in a narrow field). Peter had quite a knack for contradicting Nobel laureates. He even questioned a paper of Wolfgang Pauli (responsible for the exclusion principle that underlies chemistry) once upon a time.

It was precisely my discussion of the quantum foundations of evolution that prompted one audience member to ask me the following: 'Why are physicists less antagonistic to the idea of a God than biologists?' We all know where he is coming from, of course. One of the most prominent biologists of our time is Richard Dawkins, who is famous for his rather strongly-worded anti-religious views. The title of his bestseller, *The God Delusion*, sums up his beliefs rather succinctly. There are no equivalent physicists quite so vocally anti-religious. Most physicists do not actually believe in God, but we often use it as a metaphor for the universe (like Hawking's 'Mind of God').

Not only did this question stay with me, it allowed me to give a humorous conclusion to the whole proceedings. 'Because physicists are closer to God than biologists', I replied. Everyone laughed. Now, laughs are not exactly two-a-penny during quantum physics lectures, and it was especially heartening to end this particular one on a light note.

Let me now try to kill this joke by trying to explain it (as the American-Romanian cartoonist Saul Steinberg said: 'Trying to define humour is one of the definitions of humour'). Why are physicists closer to God?

Before Darwin (and Alfred Russell Wallace — a legend in Singapore, where I am writing these words), biology was taxonomy. Linnaeus, a 16th century Swedish botanist, physician and zoologist, classified animals according to their different characteristics — logic which we still use today. What he did not know, however, was how and why these different species existed. It was clear that there is a great deal of variation in complexity in living species, but Linnaeus probably thought that there is no correlation between how complex a species is and the time when it emerged. The fact that parts of biology and chemistry still resemble taxonomy today prompted the physicist Ernest Rutherford to say 'science is either physics or stamp collecting' (something he may have regretted later, when in 1908, he was awarded a Nobel Prize in chemistry). This is what I mean when I say that physicists are closer to God (or nature, or source, or whatever word you would replace it with). Explanations in physics are more powerful and detailed than in other sciences.

Darwin's momentous contribution was to unify biology and make it different to taxonomy by providing the logic of evolution. This logic explained how more complex creatures can naturally arise out of simpler ones. All living thing were, after Darwin, seen as part of a continuous stream of life. In this way, the gaps existing in Linnaeus's taxonomy were closed. I say Darwin, because he is the one credited with explaining natural selection. The history of science is littered with names overlooked, but few as much as Alfred Russell Wallace.

I mention Wallace, one of the pioneers of nature conservation as well as co-founder of the theory of evolution, because it is time he was recognised — and where better to give him recognition than in Singapore, where Wallace's presence is strongly felt, even now. Today, I am visiting the splendid Wallace Education Centre in Bukit Timah. One half is devoted to the Wallace Environmental Learning Lab, and the other is a 'Wallace trail' through the forest where he once collected. Although both Darwin and Wallace announced their theories at the same time, it was Darwin's publication *On The Origin of Species*, that made senior scientists take notice. Wallace's findings came through his travels in South East Asia (much of which was in Singapore), where he collected thousands of birds and insects.

Darwin's and Wallace's discoveries certainly closed some of the major gaps faced by biologists at the time. But the theory of evolution also exposed another gap. A rather large one. As the biologist with the shiny bald head told me at Will's dinner party, the biggest question in biology is 'how did living matter decide to be alive?' The gap between inanimate matter, the subject of physics and chemistry, and animate matter. How does the former give rise to the latter? How did life originate from a lifeless universe?

Darwin and Wallace explained how more complex living creatures can arise out of simpler ones by random mutations followed by the environmental selection, but neither could explain the origin of the simplest life forms themselves. Given some simple life forms, Darwin argued that the forces of evolution could, in principle, produce the rest.

The mistakes in replication of living organisms mean that every next generation will be different from previous ones. Some of their features may be more (and others less) suited to their environment. Those whose features are unsuitable will be eliminated by the environment, while those better adapted will survive. Their features will be largely imprinted into the next generation, with some mutations, and the whole story continues. The 'wrong mistakes' get eliminated, the right ones become a feature of the new genetic makeup (apart from in the case of humans, of course, as Philip Larkin's poem 'This Be The Verse' nicely illustrates). But generally speaking, we get better adapted individuals as different species evolve. Darwin provided a simple and unifying story to the whole of biology which otherwise would just be a somewhat haphazard classification of animals we happen to observe around us.

This leads us to view life as a continuous stream, from the simplest to the most complicated living organisms. It is worth remembering how recent this view really is, though it may now to us seem entirely natural and perhaps even obvious. Virgil, the Roman poet, believed that bees could be generated from the carcass of an ox. In other words, it was natural to the antiquated mind that life could be created out of no life. Similarly, as late as the eighteenth century, the prevailing belief was that maggots can somehow spontaneously emerge out of decaying meat. It took the French chemist, Louis Pasteur, to show that this is not possible, that life cannot be generated from no-life in this way: it always comes from some pre-existing life.

Another problem that was solved by the logic of evolution, albeit indirectly, is that of biological reproduction. It led us to understand that biological information is stored in an acidic molecule called deoxyribonucleic acid (DNA to you and me). When a biological entity reproduces, the first thing it does is copy its DNA. This then presents the blueprint for the new entity. This logic — ultimately based on information — avoids the infinite regression that plagued earlier philosophers.

There are two related problems with reproduction that evolution solves. First of all, in the words of von Neumann, 'if an automaton has the ability to construct another one, there must be a decrease in complication as we go from the parent to the construct. That is, if A can produce B, then A in some way must have contained a complete description of B. In this sense, it would therefore seem that a certain degenerating tendency must be expected, some decrease in complexity as one automaton makes another automaton.' This is a pretty damaging objection as it seems to completely contradict everyday experience. Life appears to be getting more and more complex, rather than simplifying into less complex organisms.

The second main objection to self-reproduction is related to the previous one, only that now it also seems to contradict logic and not just experience. If A has to make another machine B, it then seems that B needs somehow to be contained within A initially. But imagine that B wants to then reproduce into C. This means that C must have been contained in B, but since B is contained in A, C must also be contained in A. Is this ringing bells yet? So, in essence, what we are trying to say is, if we want to make sure that something lasts through hundreds of generations, it would appear that we would have to store all the subsequent copies in the initial copy. If we generalise this to an infinite number of copies, then there is clearly a resource impossibility, given that A would then have to store an infinite amount of information. The infinity problem rears its fascinating head once again.

So, as I walk down Wallace Way, it is clear that while Wallace and Darwin closed one big gap in biology, they opened up a few other ones. And this is always the case with the progress in science: we explain one mystery, only to uncover a range of other new ones.

First of all, unlike Mendeleev, Darwin did not explain why each and every one of the existing species needed to exist. This would be very hard to do, as it requires us to understand the exact environmental details throughout the history of the earth. I mean, we are still arguing about the big events that might have happened in earth's history, so we're not exactly close to understanding the entirety of earth's environmental history. Did a comet hit it? Is that what explains dino-death? Or something else? There are still quite a few gaps to fill in, although I am sure it is just a matter of time.

If Darwin could have written a periodic table of animals, this would have presumably allowed him to predict which animal species we are yet to discover (just like the six missing elements from Mendeleev's original table) or which are yet to evolve. Imagine! Well, obviously we cannot, and, in fact, hundreds of new species are being discovered each year, each discovery coming as a surprise — we cannot anticipate them given our current understanding of biology.

Predicting the future is the first gap in biology. Questions like these are hard for biology to answer, because they seem to depend on many apparent contingencies, as well as requiring us to have a stupendous amount of information about the detailed dynamics of our planet, the rest of the solar system, and beyond. We will call it the 'prediction problem'.

When you look at theories in physics, they are phrased very precisely with mathematical formulae. If you are trying to make a prediction of a physical system, you can do this extremely well and with very high precision. For example, if you want to know where Mars will be in 10,000 years' time, then physics can actually give you an answer — the laws of physics are so accurate that the answer would actually be extremely precise.

Planets have been discovered with the laws of physics — Newton's gravity. The astronomer Urbain Le Verrier mathematically predicted its existence and the exact location by looking at the motion of neighbouring Uranus and concluding that there must be another object nearby to have a gravitational effect on it. He then went to his friends in observational astronomy and told them exactly where to point their telescopes to. And so, on 24th September 1846, they saw what Verrier had predicted. This was a mind-blowing confirmation of Newtonian physics in the nineteenth century, as well as a momentous event for astrophysics.

How would biology respond to a similar question if you asked what will happen to one of our existing species (say, humans) in terms of evolution in 10,000 years? It is certainly not elementary, my dear Watson, and no one seems to have much of a clue how to make this more mathematical. There are no mathematical equations governing the laws of biological evolution. Biology cannot follow the physics paradigm: write down the equations, calculate their solutions in a particular case, do the experiment, then compare the two.

Things could be said to be even worse. We can use the logic of Turing machines and universal computers to argue that certain questions are undecidable in biology, just as they were in chemistry, providing that biological evolution can be simulated on a computer. This, of course, assumes that we may simulate biological evolution on a universal Turing machine.

Interestingly, such a simulation does seem to exist. There is a computer simulation of biological evolution called the 'Game of Life'. It was invented by a British mathematician from Princeton, John Conway, in 1970. He imagined a two-dimensional square grid, just like a chessboard, but of unlimited size. Some squares are turned on, in other words they are alive, while others are turned off, being dead. The rules of the evolution of this system are very simple:

1. Any live cell with fewer than two live neighbours dies (out of loneliness caused by under-population).
2. Any live cell with two or three live neighbours lives on to the next generation.
3. Any live cell with more than three live neighbours dies, (being depressed by over-population).
4. Any dead cell with exactly three live neighbours becomes a live cell (meant to simulate some kind of reproduction).

A remarkable point about the 'Game of Life' is that it gives rise to complex patterns that move about, interact with other patters and then perish. In other words, they live, explore the environment, multiply and then ultimately die, just like all living systems. Most initial configurations of alive and dead squares actually lead to all squares being dead — nothing living ever gets created with these starting configurations. Other configurations give rise to permanent patterns, which not only persist, but go on to multiply, creating copies of their own. They enter a complex pattern of interactions whose future is very much full of complexity.

The most remarkable thing about Conway's system is exactly its unpredictability; it is in fact a universal computer. It can do whatever any other computer (like your laptop, desktop, iPad) can. This is as good as it gets for a machine, and yet it carries its own limitations with it. There is no way to predict for a generic initial state of the grid if anything exciting will appear at some point in time and persist.

If real biological evolution is like this (a big 'if') then even if we had a perfect mathematical understanding of its dynamics, it would follow that we can never predict if and when a given species will appear or go extinct.

This would be no surprise to my colleague Mile Gu, whom I mentioned earlier. He showed that there are things we cannot predict in chemistry as this would be tantamount to solving the Halting problem. So, in biology, which is even more complicated, we definitely expect the same to be the case.

The big 'if' refers to two things. First of all, it is not clear if the Game of Life captures evolution properly. I mean, we do not know if there is more than that to evolution, for a start. We find it even hard to define rigorously what it means to be alive. Secondly, maybe the real evolution started from a state from which it is easier to draw some conclusions about the future. Maybe it is completely deterministic (like the laws of Newtonian physics), but we just have not figured out how to predict the biological future. We simply do not know.

But, actually, biology can do better than stop at the complexity of the Game of Life. The first time I saw how far you can go in biology as far as reductionism was concerned was age 18, when I came across Richard Dawkin's *The Selfish Gene*. It became an instant international best-seller — for good reason. Not only is it a very lucid update on Darwin's theory, but this book was the first time I saw someone try to make the theory of evolution mathematical, to explain it fully, and to make predictions based on it.

And that is the point. If evolution is going to be a proper scientific theory, we know that it must be falsifiable. There must be some experiments, even if only in principle, that would invalidate Darwin. British scientist J.B.S. Haldane, when asked what would constitute evidence against evolution, famously said, 'Fossil rabbits in the Precambrian.' The distribution of fossils on earth in space and time reflects the growing complexity generated in time by the evolutionary process of random mutations and natural selection. Finding ape fossils deeper than any dinosaur's is something that ought not to happen unless some prankster put it there deliberately. Note that the falsification of biology would still come from the past. It is not a prediction that is invalidated, but what philosophers call a retrodiction. Of course, you can make a retrodiction into a prediction by predicting that no one will find fossils of rabbits in the Precambrian in the future!

Dawkins discussed how far we can go down in terms of simplicity, in trying to explain everything in the biological world in terms of very simple units — in this case, of course, these units are genes. Dawkins was writing from a very different viewpoint, as of course it was a century after Darwin and Wallace's discoveries that genes became a way to explain evolution.

This explanation of biology using mathematics made me see that the methodology of physics could ultimately be applied to biology and really reduce it even more. This is because we know that once you reduce biological behaviour to genetics, then of course you now are working with genetics and molecules, and that is the subject of chemistry, which itself is grounded in quantum physics. So, in a way, you have got this beautiful pyramid of explanations: starting from quantum physics, then explaining basic chemical laws based on quantum physics and then from chemistry, we try to explain genetics and then more complicated living organisms. Somehow the whole fits this nice scientific logic.

How much of biology is really just physics? Could it be that some phenomena that look complex in biology are really just consequences of some simpler chemistry and ultimately, even simpler quantum physics?

Evolution, as it is normally understood, is based on two principles; we have random mutations that lead to changes in the new generations. These are then eliminated by the environment, a process known as natural selection. Elimination simply means that the individuals die before they managed to reproduce. Both of these processes, random mutations and natural selection, look simple enough, however, on closer inspection, they raise many questions.

Take the phrase 'random'. To a biologist, it does not mean the same as what a physicist considers random (as in quantum physics, where it means that there are no underlying causes for something to happen). In biology, it means that the mutation does not have a foresight, in the sense that it is not engineered to make something deliberately different and designed for future function. When the mutation happened that made our predecessors' skull softer at the top, it seemed like a detrimental feature as the skull protects the brain. However, the softer skull also allowed our brains to enlarge and this, in the long run, had a positive effect on human evolution (in fact, strictly speaking, we do not really know if the effect will ultimately be positive for us, as larger brains lead to humans making better weapons, a feature that might ultimately lead to our extinction! But, well, it does illustrate my point that mutations have no foresight).

On the other hand, mutations are far from random; they are constrained by the laws of chemistry and, as we argued, ultimately by the laws of physics. When genes mutate, they do so one small feature at a time. A couple of protons gets dislodged and leads to a different kind of information copying. One molecule in the DNA, instead of pairing up with another, ends up pairing up with a different molecule. But the choices are limited and constrained by the structure that has evolved over billions of years.

On top of it, the information genes carry has to be transferred to the individual's phenotype, i.e. to their form and function. The genes might instruct the growth of an arm or a leg, but this growth itself is also constrained by the laws of physics (not biology). The form and function in turn is what the environment acts on, but this is not a one-way process. Individuals also alter their environment, which in turn alters them and so on.

The first time the role of physics in biology was beautifully explored and presented was by D'Arcy Wentworth Thompson, a Scottish biologist, mathematician and classicist. Legend has it that he was so erudite that when he interviewed for a university post they offered it to him in all three disciplines. One could argue that he was one of the last Renaissance men.

Wentworth Thompson's 1917 thesis in *On Growth and Form* was that biologists ignore the role of physics to their detriment. According to him, physics constrains biological form so strongly, that there is in fact very little room for manoeuvre. There are a few of my favourite examples of Wentworth Thompson's, which demonstrate that genes cannot do anything they please.

Why is the biggest land animal the elephant while the biggest water one is the whale? Are their overall features an accident and why are both mammals (i.e. warm blooded)? One could easily revert to the lazy answer: their genes dictated this. But this answer would be a cop out. We can do much better.

In fact, the first person to explain this and related features was a physicist (biology did not even exist then, really) called Galileo Galilei. The answer is: gravity. Genes cannot do anything they want. For instance, they cannot create an animal that is 100 meters tall. The explanation for this is simple. As any animal becomes larger, its weight varies in relation to its unit length cubed (in other words, in proportion to its volume), while the area (and hence the strength) of its legs varies in relationship to its unit length squared. And this puts a huge pressure on its support.

If the animal doubles in every direction, gravity becomes eight times stronger and so does the pressure on its bones. This scale effect sets a boundary on the upper scale of all life in our gravity since a large animal's legs would eventually break due to its own weight. This explains why elephants (and rhinos and hippos) have thick legs.

Whales, on the other hand, live in water and gravity is not so relevant there. They can therefore afford to be bigger (and their shape is more aerodynamic, or hydrodynamic, of course). Not only that, but whales are mammals and need to maintain their own body temperature in the frequently cold deep oceanic waters. For that, they need lots of muscle, i.e. a large body weight. Since the temperature is lost through the surface (which scales as the square of the length), whales actually gain by being larger as mass grows faster as unit length cubed.

We can apply similar logic to explain why mice are the smallest mammals. This is limited by the amount of food the mouse needs to take in order to maintain its temperature. Or, to throw another seemingly unrelated question: why some animals of very different masses can more or less jump to the same height, independently of their size (both a cricket and a human can jump about a meter or so vertically)? This is to do with the potential energy of jump being proportional to mass times height, while the muscular work strength is proportional to the mass (equating the two gives us that height is the same independently of the mass). The fact that all animals can jump to more or less the same height is not because we have the same gene for jumping, but it is because the force of gravity equals m times g (that is mass times the Newtonian gravitational constant *g*).

A similar logic works for the many different species of birds and insects that I had the pleasure of seeing earlier today on my Wallace walk. It is not just the fact that aeroplanes can fly that demonstrates the existence of atoms and molecules in the air. Birds and insects, of course, do not use petrol to propel themselves forward — they actually need to flap their wings. The question is if we can use some simple physics to tell how frequently a bird needs to flap its wings in order to be able to maintain itself in the air. The buoyancy pressure due to flapping has to be as large as the gravity pulling the bird down. Now, there are only a handful of factors that determine this condition: the mass of the bird, the size (area) of its wings, the span of the wings, the density of air and the frequency of the bird's flaps. There is actually only one way of putting these quantities consistently into an equation so that it makes logical sense. However, the point is that physics puts constraints on the size of the bird and the span of the wings and the mass of the bird and so on, in order for it to be able to be buoyant. For instance, we know that the frequency of flaps is inversely proportional to the mass of the bird. Smaller birds have to move their wings faster in order to stay afloat. And poor bees work overtime.

This of course illustrates the point that we cannot just treat physics and biology as isolated disciplines. Physics clearly has something to say about biology at every level. But — does the causation (or, can the causation) go the other way? Does biology constrain physics too? I will go more into this later.

It is undeniably true that the biological reduction to chemistry and then physics makes us feel more uncomfortable than the reduction of chemistry to quantum physics. This is because as living systems, we believe that we have a high degree of self-determination and autonomy. Reducing biology to (quantum) physics might suggest to us that our behaviour — and that of the rest of living systems — is somehow determined by minute atomic interactions (which might ultimately be random!). Where does that leave our feeling that we are in charge of ourselves?

Some scientists are comfortable with determinism. Here is what Victorian biologist Thomas Henry Huxley had to say in his essay 'On the Hypothesis that Animals are Automata': 'What proof is there that brutes are other than a superior race of marionettes? Which eat without pleasure, cry without pain, desire nothing, know nothing, and only simulate intelligence?'

We have no more evidence to the contrary even today. Nothing to prove that we are more than embodiments of the laws of physics. Maybe we just think we can control more than we do, and can somehow act contrary to nature. There is a story — some say originating with Bertrand Russell — about a man who daily, first thing in the morning, stood on one foot for ten minutes. When challenged to explain why he did that, he replied 'well, it keeps tigers away, of course'. But there are no tigers in England, his friends protested. "You see! It works," the guy explained.

Are we all like this man? I mean, do we just think we control more stuff around us than we actually do? Maybe we feel we act freely, but do we? I'm sitting outside my favourite bar in Clarke Quay, a good place to ponder. The sky is heavy and the Quay, as always after dark, crammed with people. It is clean and beautiful, but unlike Beijing, this is not a microcosm of clean: it is no different to the rest of Singapore. A century ago, this was almost all native forest. Some words of Wallace's I read earlier are playing round in my head. 'Future ages will certainly look back upon us as a people so immersed in the pursuit of wealth as to be blind to higher considerations', he wrote, back in 1863. Is it blindness, or it our desire to be in control?

I am reminded of placebo buttons, a concept which at first made me laugh, then despair. You know when you are in a lift and you press the button to close the door faster? Well, in most lifts, these buttons do not actually do anything. I mean, they are not connected to anything! They are there just to pacify you and make you feel in charge. The same is apparently true for many traffic lights.

Could our feeling of being in charge and not pre-determined be a naturally-engineered placebo switch? I think this is a fascinating possibility, though it is hard to see how to test this hypothesis in practice. How would one show that when we switch on the feeling of being in charge, this is actually not connected to anything else in our system?

It is very hard to imagine what kind of experiment would demonstrate that we act freely. Suppose we decide to act contrary to our impulse. Does that prove anything? Not really, since we could be determined to act against our impulse. The German poet and polymath, Goethe, said of nature: 'We obey her laws even when we rebel against them; we work with her even when we decide to act against her'.

Maybe by trying to reduce biology to physics, we are actually committing the fallacy of confusing the cause and the effect. Nassim Taleb, the bestselling author of *The Black Swan*, tells the following story. He wanted to develop his muscles more and acquire a pleasing body-shape. But he did not want to go to the gym like the majority ('too boring') or play tennis ('too middle class'). So, he opted for swimming. After all, look at a random professional water polo team. They all have fantastic bodies.

However, after one year of regular swimming, Taleb realised something disappointing. He was no doubt fitter, but his body shape did not improve. The reason? He confused the cause and the effect. It is not that swimming leads to a better body, but quite the opposite — those with great bodies are actually better swimmers!

The same goes for beauty product commercials, of course. Women do not become better looking by using them; on the contrary, those who are selected to use them in advertisements are already good looking.

With this in mind, which way does the causation work as far as physics and biology are concerned?

Many biologists believe not only that biology is not reducible to physics, but that biology is prior to physics, in which case, we are mixing the cause and effect. In other words, we should be trying to understand physics from the biological perspective and not the other way around!

Maybe this is not that unusual. After all, doing science is a uniquely human trait. This is our way — or one of them — of understanding the world. No other species could understand quantum physics or any other aspect of science. This means that the biological evolution had to reach a certain level of complexity for organisms like us to emerge with the ability to do physics. In that sense, the biological evolution comes before physics. Putting it more strongly, biology is actually — in this view — necessary for physics.

This is all undoubtedly true, but it would be weird to think that the laws of physics only became functional once humans have evolved. We know the universe behaved in the same physically ordered way before humans existed, and even well before life emerged some four billion years ago. Also, even if humans and all life went extinct, the laws of physics would still presumably be valid and the universe would obey them (though there might be no one around to confirm that). The way we express the regularities in nature as mathematical laws may well be uniquely human (though we do not know what some advanced aliens might make of it), but the fact that the universe is lawful is beyond doubt.

Reduction of physics to biology also carries a psychological penalty. We believe that intentionality, namely that biological systems do things deliberately to their environment, and unlike inanimate matter, is something that is intrinsically human. We are responsible for what we do because we decide to do it following our intentions.

Living systems, according to this logic, are not computers.

Even some prominent physicists, such as fellow Oxonian Roger Penrose, pursue this intuition to claim that the human consciousness is more powerful than computers. Penrose claims that the undecidability, and the Halting problem, whilst clearly are limitations for a computer, are actually not obstacles for a conscious system. The brain not being a computer is a perfectly respectable scientific hypothesis to entertain. We simply do not know. We do not even know if the rest of biology, without conscious organisms (whatever they are), can actually be faithfully simulated on a computer. (Faithfully here means 'while capturing all the important details'.)

And yet, the undeniable fact is this. Everything is made up of atoms. Their existence and structure is fully captured by quantum physics. All living systems, plants and animals are also made up of atoms. These interact, forming more complicated molecular structures. Resulting chemical processes actually make the living things living (though, admittedly, we do not quite understand how). Therefore, biology should actually ultimately be understood in terms of physics. Living matter obeys laws of physics and should be understood using them. The gap between physics and biology ought to be reducible.

The first scientist to take this seriously was the physicist Erwin Schrödinger. This is the same person who discovered the laws of quantum physics, and most likely it is not an accident that he also contemplated the reduction of biology to physics.

CHAPTER IV

UNITING THE NATURAL SCIENCES

After revolutionising physics, Erwin Schrödinger turned to biology. In 1944, he wrote a highly influential book, *What is Life?*, that discussed the physical basis of biological processes. From where we're standing now, he got an astonishing number of things right. He basically (I say basically...) anticipated the stable encoding of biological genetic information, and he guessed that crystals form the basis for encoding (Watson and Crick later showed DNA has a periodic crystalline-like structure). He also understood the thermodynamical underpinnings of living processes that have since become the bread and butter of biology.

The most frequently quoted paragraph from *What is Life?* is:

'...*living matter, while not eluding the "laws of physics" as established up to date, is likely to involve "other laws of physics" hitherto unknown, which however, once they have been revealed, will form just as integral a part of science as the former.*'

The paragraph has been interpreted in two different ways. One is that the other laws of physics are those of quantum mechanics. What he is then saying is that life could not be understood using just the laws of classical physics. From what we have seen so far, this is undoubtedly true. Classical physics cannot even explain atoms.

But maybe Schrödinger is saying that even quantum physics is not enough. This would be a radical view but also possible, given that the book was written 20 years after the discovery of quantum physics. I heard this view echoed recently by Galen Strawson:

'It is beyond serious doubt that conscious experience is wholly a matter of brain activity, but this doesn't show us that we don't know what consciousness is. It shows us that we don't know what matter is. The hard problem is the problem of matter. Matter is even more extraordinary than we thought, as physicists have been demonstrating for a long time.'

Strawson agrees that consciousness will be reducible to the activity of the brain matter, but what this suggests to us is that we do not really understand what matter is. Maybe we really need new laws of physics? Similar amazement was expressed at the difference between living and non-living matter by the American naturalist Joseph Wood Krutch: 'No wonder that enthusiastic biologists in the nineteenth century, anxious to conclude that there was no qualitative difference between life and chemical processes, tried to believe that the crystal furnished the link, that its growth was actually the same as the growth of a living organism. No one, I think, believes anything of the sort today. Protoplasm is a colloid and the colloids are fundamentally different from the crystalline substances. Instead of crystallising, they jell, and life in its simplest known form is a shapeless blob of rebellious jelly rather than a crystal eternally obeying the most ancient law.' To Krutch, life is a blob of rebellious jelly and very distinct from inanimate matter.

And here lies our big gaping hole in our Great Reduction of the natural sciences. The first part of this: How does inanimate matter lead to life? And the second, intrinsically linked: How does life become conscious? Let us go on our final journey into the natural sciences. We will start quantum: can quantum physics help us to answer these questions? I am beginning this journey about 40,000 feet above the earth's surface. It is early in the morning and clear out, and the Bay of Bengal shimmers way in the distance. It is a good place to think of the bigger picture from.

QUANTUM NATURE

When Schrödinger began his treatment of biology using the laws of physics, he followed in the footsteps of Ludwig Boltzmann. As a refresher, Boltzmann's theory started with how the first and the second law of thermodynamics drive biological processes. The first law stipulates that the overall energy has to be conserved in all processes, though it can transform from heat (a disordered form of energy) to work (a useful form of energy). The second law says that the overall disorder (as quantified by entropy) has to increase in a closed system, but of course, this does not prohibit one part of the system from becoming more ordered at the expense of the rest, which becomes even more disordered. The key is how to get as much useful energy (work) as possible within the constraints of the second law, namely that the overall disorder has to increase. The trick is to make the rest even more disordered, i.e. of higher entropy, and then to exploit the entropy difference. This is the trick that life pulls.

Boltzmann expressed this thermodynamically-driven logic beauti-fully in the late nineteenth century when he said: "The general struggle for existence of living beings is therefore not a fight for energy, which is plentiful in the form of heat, unfortunately untransformably, in every body. Rather, it is a struggle for entropy that becomes available through the flow of energy from the hot sun to the cold earth. To make the fullest use of this energy, the plants spread out the immeasurable areas of their leaves and harness the sun's energy by a process as yet unexplored, before it sinks down to the temperature level of our earth, to drive chemical syntheses of which one has no inkling as yet in our laboratories."

Here is what Schrödinger had to say about it: '... the laws of physics, as we know them, are statistical laws. They have a lot to do with the natural tendency of things to go over into disorder. But, to reconcile the high durability of the hereditary substance with its minute size, we had to evade the tendency to disorder by "inventing the molecule", in fact, an unusually large molecule, which has to be a masterpiece of highly differentiated order, safeguarded by the conjuring rod of quantum theory. The laws of chance are not invalidated by this "invention", but their outcome is modified.'

It is a beautiful idea. We cannot change the fact that atoms are random, but we can minimise the effect of randomness by making larger and larger structures. There is more fault-tolerance in larger structures, safety in numbers. And quantum physics is the key, prolonging stability amid the descent into decay dictated by the second law.

According to this view, all living systems are actually Maxwell's 'demons'. This is not in the original sense that Maxwell meant, namely that demons can violate the second law, but in the sense that we all try to maximise the information about energy and utilise it to minimise entropy and therefore, extract useful work. This view of life is beautifully articulated by the French biologist and Nobel laureate Jacques Monod in his classic *Chance and Necessity*. Monod even goes on to say: 'It is legitimate to view the irreversibility of evolution as an expression of the second law in the biosphere.' In animals, the key energy-generating processes take place in the mitochondria, which converts food into useful energy; plants rely on photosynthesis instead of food. We can almost take the fact that living systems 'strive' to convert heat into useful work as a defining feature that discriminates life from inanimate matter.

There are, of course, always grey areas. Man-made machines, like cars, also convert energy (fuel) into work, but they do not really strive to do it independently of us. On the other hand, we ourselves are not independent of the external factors either, and so the whole issue regarding how to define life is not that easy.

This is why biologists tend to think about all living processes as driven by entropic forces. This force is fictitious (i.e. not fundamental, like the electromagnetic one) and it captures the fact that life maintains itself at an entropy value lower than that of the environment. The entropy gradient that drives all life is, of course, ultimately based on the temperature difference between the sun and the earth.

Another physicist, Leon Brillouin, called this the neg-entropy ('neg' being short for negative) principle. Life operates far from equilibrium with its environment, which is characterised by the maximum entropy. To maintain itself far from equilibrium, life needs to import negative entropy from the environment. This is why we eat food which is highly structured, either in the form of plants (which use the earth-sun neg-entropy to develop their own structure through processes involving carbon dioxide and water) or animals (that eat plants to maintain low entropy). We utilise the chemical energy stored in the bonds between atoms that make up our food.

The neg-entropy provided by the sun-earth temperature difference can be estimated to be 10 to the power of 37 times Boltzmann constant per second. This is a huge amount. How much neg-entropy is required to create life? Well, we can assume that to turn lifeless earth into present earth, it is necessary to pluck every atom required for the biomass from the atmosphere and place it into its exact present-day quantum state. These assumptions maximise the entropy of dead-earth and minimise that of earth, meaning the difference between the two entropies grossly overestimates the entropy reduction needed for life. A simple calculation shows that the entropy difference required is about 10 to the power of 44 times Boltzmann constant. This suggests that in principle, about an hour (10 to the power of 44 − 37 = 7) of sun-earth motion should give us enough entropy to create all life!

As we know, it took much longer for life to evolve — partly because the conditions were not right, and also because the whole process is random. Living systems are by no means perfect work engines, and so on, but the bottom line is that the sun-earth temperature difference is enough to maintain life. On top of this, we have all the entropy generated by life, such as culture, industry, big cities and so on, things of which can help us handle energy more efficiently.

Anything that would require more than 10 to 53 units of entropy (a unit of entropy is k times T where k is Boltzmann's constant and T the temperature) would take longer than the age of universe. The particular way in which organisms on earth extract energy puts more severe limitations on the sustainability of life. For instance, it is only plants that utilise the sun's energy directly. Herbivores obtain it by eating plants and carnivores by eating herbivores (and plants). The more removed an organism is from direct sunlight utilisation, the less efficient its extraction. Isaac Asimov in *Life and Energy* estimated that the sun-earth system can sustain at best 1.5 trillion humans (eating algae directly!). We are still far away from this, but the bound assumes that we have the relevant technology to do this too.

The first person to take Schrödinger's suggestion that life needs quantum physics seriously was a Swedish condensed matter physicist Per-Olov Löwdin in 1964. He was actually the one to coin the phrase 'quantum biology' when writing about the mechanisms behind DNA mutations. The main quantum contribution he had in mind was proton tunnelling and its role in DNA replication and genetic mutations. It was already known that the explanation of chemical bonding itself lies in quantum physics — but the tunnelling was an additional feature.

Tunnelling is a quantum effect where a particle manages to go through a barrier which classically it could not possibly overcome due to insufficient energy. A typical example is throwing a tennis ball at a hard wall. It always bounces back. However, in quantum physics, there is always a chance that the ball will actually go through. Nuclear decay is explained this way. The protons and neutrons that comprise an atomic nucleus can actually escape it (even though they do not have enough energy to do it) thereby leading to nuclear decay.

If tunnelling really is fundamental to DNA replication and mutations, then of course quantum physics is crucial for life (as tunnelling cannot occur in classical physics). The key in DNA replication is the matching between different base pairs on two different strands of the DNA and the position of protons required for base pairing. If protons tunnel to a different spatial location, then a mismatch in the pairs can occur, which effectively constitutes a genetic mutation.

The main idea behind biological quantum tunnelling is that a certain process can still take place in quantum physics even if there is not enough energy for it under classical physics. So, even if the energy due to temperature is not sufficient to make certain molecules pair up, quantum physics tells us that there is a chance a process can still take place. The only slight flaw in this is that almost 60 years have passed since Löwdin's proposal, but no conclusive evidence is available to prove that protons tunnel in DNA and that this has an impact on DNA replication. Something tells us, however, it is just a matter of time before it is. Before it becomes available, though, there is another form of evidence that quantum physics might be at work in other biological phenomena.

Photosynthesis is the name of the mechanism by which plants absorb, store, and use light energy from the sun. Solar energy is used to utilise carbon dioxide and water to build various plant structures such as roots, branches and leaves. Recent fascinating experiments led by Graham Fleming at the University of Berkeley, California, suggest that quantum effects might matter in photosynthesis. Furthermore, they point out a close connection between the photosynthetic energy transfer and certain types of quantum computations. In other words, plants are so much more efficient than what we expected, that there must be some underlying quantum information processing.

Plants are fiendishly efficient (between 90–99%) at channelling the light absorbed by antennas in their leaves to energy storage sites. The best man-made photocells barely achieve a 20% efficiency.

So, there is an enormous difference — but how do plants do it? The complete answer is not entirely clear, but the overall picture is this: when sunlight hits a surface that is not designed to carefully absorb and store it, the energy is usually dissipated to heat within the surface. Either way, it is lost as far as any subsequent useful work is concerned. The dissipation within the surface happens because each atom in the surface acts independently of other atoms.

When radiation is absorbed in this incoherent way, then all its useful properties vanish. What is needed is that atoms and molecules in the surface act in unison. And this is a feat that all green plants manage to achieve. In order to understand how this happens, it is helpful to think of each molecule as a small vibrating string. All molecules vibrate as they interact with one another, transferring energy between them. When they are hit by light, they change their vibration and dynamics and end up in the most stable configuration. The crux is that if the vibrations are not quantum, then they cannot find the reliable configuration as efficiently (at best with 50% efficiency).

Fleming's experiments were initially performed at low temperature (77 Kelvin, while plants normally operate at 300 Kelvin), but the subsequent studies indicated this behaviour persists at room temperature (although to date, the experiments have not been done using sunlight). Therefore, it is not entirely clear if any quantum effects can survive under fully realistic conditions. However, even the fact that there exists a real possibility that a quantum computation has been implemented by living systems is a very exciting and growing area of research. To explain, in the case of photosynthesis, the information that is conveyed is simply the energy of photon and the vibrations are a form of quantum computation that transfers this information to the appropriate reaction centre where chemistry takes over to generate energy.

Magneto-reception is the other instance where animals might be utilising quantum physics. European robins are crafty little creatures. Each year they make a round trip from the cold Scandinavian peninsula to the warm equatorial planes of Africa, a hazardous trip of about four thousand miles each way. Armed with only their internal sense of direction, these diligent birds make the annual journey without any fuss (that we know of).

When faced with a similar challenge, what do we humans do? Ferdinand Magellan (the late 15th-century Portuguese explorer), for instance, had the same problem — and solved it. He worked out what a useful contribution a compass could make to your journey: he showed how the earth's magnetic field — to which the compass is sensitive — could be used as a stable reference system to circumnavigate the earth. So now, when we start in Europe and use a compass to follow the southward direction of the earth's magnetic field, we are confident we will eventually find ourselves in Africa. But while a compass may guide us humans, it is not at all clear how robins find their way so unerringly and consistently. Do they also have a kind of inbuilt compass? I mean, it only makes sense that there is some kind of internal guidance mechanism, but it is certainly not of the type Magellan used.

German biologist Wolfgang Wiltschko came up with the first evidence of this guidance mechanism in the early 1970s. He caught robins on their flight path from Scandinavia to Africa and put them in different artificial magnetic fields to test their behaviour (be assured, no harm was done to the robins!). One of Wiltschko's key insights was to interchange the direction of the north and south and then observe how the robins reacted to this. Much to his surprise, nothing happened! Robins simply did not notice the reversal of the magnetic field. This is very telling: if you did this swap with a compass, its needle would follow the external field, make a U-turn, and point in the opposite direction to its original one. The human traveller would be totally confused. But somehow, the robins proved to be impermeable to the change. Huh?

Wiltschko's experiments went on to show that although robins cannot tell magnetic north from magnetic south, they are able to estimate the *angle* the magnetic field makes with earth's surface. And this is all they needed in order to navigate themselves. Incredible.

In a separate experiment, Wiltschko covered robins' eyes (again, no harm, I promise) and discovered that they were unable to detect the magnetic field at all (come to think of it — how on earth do you cover robins' eyes?). He concluded that, without light, the robins cannot even 'see' the magnetic field, whereas of course a compass works perfectly well in the dark. This was a significant breakthrough in our understanding of the birds' navigation mechanism. The now widely-accepted explanation of Wiltschko's result was proposed by Klaus Schulten and developed by Thorsten Ritz, biologists both studying in America.

The main idea behind the proposal is that light excites electrons in the molecules of the retina of the robin. The key, however, is that the excitation also causes the electron to become 'super-correlated' with another electron in the same molecule. This super-correlation, which is a purely quantum mechanical effect, manifests itself in the form that whatever is happening to one electron somehow affects the other — they have become inseparable 'twins'. Given that each of these twinned electrons is under the influence of the earth's magnetic field, the field can be adjusted to affect the relative degree of 'super-correlation'. So, by picking up on the relative degree of 'super-correlation' (and relating this to the variation in the magnetic field) the birds somehow form an image of the magnetic field in their mind, and then use this to orient and navigate themselves. As a physicist, Ritz already knew a great deal about this super-correlation phenomenon: it had been proven many times in quantum physics under the name of 'quantum entanglement'.

Our very simple model suggests that the computation performed by these robins is as powerful (in the sense that entanglement lasts longer) as any similar quantum computation we can currently perform! More specifically, robins can keep electrons entangled up to 100 microseconds, while we humans can manage just about the same (at room temperature). If this is corroborated by further evidence, its implications would be truly remarkable. For one, this would make quantum computation yet another technology discovered by nature long before any of us humans thought it possible. While nature continues to humble us, it also brings new hope: the realisation of a large-scale usable quantum computer is possibly not as distant as we once thought. All we need to do is perhaps find a way of better replicating what already exists out there in the natural world.

In 1932, Niels Bohr, one of the fathers of quantum physics, gave a series of lectures entitled "Light and life". In these lectures, he discussed, among many other things, the potential relevance of quantum physics for biology. One of the often-quoted paragraphs is the analogy he drew between the existence of quanta in physics — which is a simple brute fact of quantum physics that does not seem to have a deeper underlying logic, and the existence of living systems in biology (which he thought might also be a fact we just have to accept at its face value). He said:

"... the existence of life must be considered as an elementary fact that cannot be explained, but must be taken as a starting point in biology, in a similar way as the quantum of action, which appears as an irrational element from the point of view of the classical mechanical physics, taken together with the existence of elementary particles, forms the foundation of atomic physics. The asserted impossibility of a physical or chemical explanation of the function peculiar to life would in this sense be analogous to the insufficiency of the mechanical analysis for the understanding of the stability of atoms."

What Bohr seems to be asserting is, that we cannot explain the emergence of living systems out of non-living ones by using (just) the laws of physics (and hence, chemistry) in the same way that we cannot really explain why the world is quantum (*a priori*, it could have been governed by any other laws of physics). In this sense, one would say that Bohr was clearly a non-reductionist.

Interestingly, Bohr does not consider the possibility that some new laws of physics could actually be able to explain how living matter arises out of non-living. This was, for instance, entertained by, among others, the physicist Rudolf Peierls. When asked if physics will be able to explain life, he said: "I don't think anything terribly mysterious is to be expected here; it is rather like what happened to physics in the nineteenth century when, at first, scientists believed that any explanation had to involve a mechanism...When they met electric and magnetic phenomena, physicists tried to explain these with some kind of mechanism. Maxwell even tried to do that, but then he realized, and other people realized, that this didn't make sense because electricity and magnetism were physical concepts in their own right, not contradicting, but adding to and enriching mechanics. And in that sense, I think we won't have finished with the fundamentals of biology until we have enriched our knowledge of physics with some new concepts."

Peierls also draws the analogy with mechanics which is not able to explain the electromagnetic phenomena (just as Bohr does), but he concludes that the new physics could be very different to the present — just like the electromagnetic phenomena were found to be different to the mechanical ones — and could therefore, naturally accommodate and explain life. The concept of the electromagnetic field cannot be found in Newton's mechanics and had to be added to it, thereby enriching the physics. The same way, Peierls thought, some new physics could also contain concept that will be enriching, in fact so much so, that they might naturally absorb understanding of life.

However, Bohr had deeper reservations about reductionism and this was based on his philosophy of complementarity. Bohr believed that much as in quantum physics, we cannot simultaneously show the wave and the particle nature of a physical object "...so might the peculiarity of the phenomena of life, and in particular the self-stabilizing power of organisms, be inseparably connected with the fundamental impossibility of a detailed analysis of the physical conditions under which life takes place".

In other words, understanding a living system physically is actually complementary to it being alive. We can either confirm that an object is alive or probe it to understand it physically (thereby killing it!), but we cannot do both.

Recently, colleagues of mine and I thought about a way to test this idea. The question we asked is if we can actually confirm that a biological system is alive while at the same time knowing that it is also in a quantum superposition. The superposition in question is between the living system and a single particle of light, a photon (very appropriate given that Bohr's lecture title was "Light and Life").

The experiment was done by my friend Dave Coles at Sheffield and in collaboration with other researchers at Oxford, Sheffield and Harvard. It consists of placing a tiny sample of the purple bacteria (that survive by photosynthesising in a quantum way described earlier) into a very small cavity (a millionth of a meter in size) and letting it interact with the photon trapped inside the cavity (the walls of the cavity are highly reflective mirrors intended to trap the photon inside for as long as needed).

The interaction between bacteria and light leads to a state in which they can no longer be discriminated from one another. This state is such that the bacteria has absorbed and not absorbed the photon at the same time and the photon exists inside the cavity and does not at the same time. In other words, it is an entangled state between bacteria and light.

The crux is that while in this quantum state, the bacteria is still alive. How do we know that? We know that because we can add another molecule that shines (so-called "dye") into the cavity and this molecule is kept out of the bacteria if they are alive. If the bacteria dies, their defence mechanism breaks and the dye enters it without a problem.

What Dave Coles found is that while the bacteria are superposed with a photon, they are also alive. Therefore, a biological system can be alive and in a quantum superposition!

However, it should be said that Bohr's statements are sufficiently vague that they could be — and since 1932 have been — interpreted in many different ways, leading to people holding widely diverging views on physics and biology.

One view, which seems opposite to Bohr's, is that maybe the mystery of life and the mystery of the origin of quantum are one and the same. This view would say that the analogy Bohr was making between quanta and life is actually more than just an analogy. Maybe quantum physics is actually necessary for the existence of life.

This hypothesis is hard to test (we would have to show that life can never take off in the classical world and how on Earth does one do that?). But we can do something else. In an interesting variant of Dave's experiment, we can actually alter the chemistry of bacteria in such a way that it can only live (by doing photosynthesis) when it is in an entangled state with light in a cavity. In this case, we can literally engineer it so that it cannot survive outside of the cavity, i.e. without quantum physics. While in quantum superpositions, these bacteria would be able to photosynthesise and live; otherwise, they would die.

It is not clear what this means as far as reducing biology to physics. But one thing is for sure. There is a lot of mileage we will get by continuing to apply physics to biology and given the current trends, we really ought to be optimistic that the gap between living and inanimate matter will surely be closed sooner or later.

A fascinating possibility is now emerging in which we try to reverse-engineer living processes and simulate them on computers. This is not just like in John Conway's *Game of Life*, where a cellular automata algorithm generates replicating patterns on the computer screen based on a simple underlying algorithm (whose code is only a few lines long). Instead, we can now start to think about simulating biological processes, such as energy flow in greater detail using the new quantum technology. After all, if living systems are those that strive to get useful work out of heat, then the *Game of Life* misses the point: none of its structures can even be thought of or become alive! Do we need to breathe in the 'striving' component into inanimate structures, or will a faithful simulation of life automatically be living? We are getting very close to our biggest gap in the Great Reduction here.

We are now already using quantum computers of up to 20 qubits to simulate and understand properties of solids. But solids are inanimate. Could it be that as our simulations become more and more intricate, we could faithfully simulate living processes with quantum computers to recreate life? And could this artificial, computational simulation really become alive? Oxford University's newest research college, the Oxford Martin School, is currently funding the first steps towards such a program and aims to develop future quantum technologies inspired by the design of natural systems. It is exciting stuff.

When Schrödinger mentioned the 'other laws of physics', could that have meant that we may need to go beyond quantum physics to understand life? Could indeed biology teach us something about the underlying laws of physics? This idea is radical, I know, and most physicists would probably reject it. After all, we think of physics as underlying chemistry and chemistry underlying biology. The arrow of causation works from small (physical systems) to large (biological) systems. But could biology tell us something about physics?

David Deutsch's 'constructor theory' could be a novel approach to physics motivated by biology which is all about constructing things. In physics, on the other hand, we talk about laws of change and states of systems. Should the laws of physics actually be expressed as some kind of construction laws akin to the laws of evolution?

The question is whether we can express the whole of physics simply by stipulating which processes are possible and which are not. This is very different from how physics is usually phrased, in both the classical and quantum regimes, in terms of states of systems and equations that describe how those states change in time. The blind alleys down which this approach can lead are easiest to understand in classical physics, where the dynamical equations we derive allow a whole host of processes that patently do not occur — the ones we have to conjure up the laws of thermodynamics expressly to forbid. But this is even more so in quantum physics where modifications of the basic laws lead to even greater contradictions.

Rephrasing the laws of physics in terms of what constructions are possible and which are not — a paradigm shift suggested by David Deutsch and then developed by Chiara Marletto also at Oxford — might give us a different perspective on bridging the physics and biology gap. After all, living systems are constructors and so is the theory of evolution. Perhaps phrasing physics in terms of construction tasks (and not in terms of dynamical laws) will make it better suited to explaining life.

In physics, we study simple, inanimate objects, so physicists find it very difficult to understand, for example, weather patterns, or financial markets. Anything that is more complicated, it seems that we do not have the same grasp that we have with atoms or things like that. And why is that? Because there is a small matter of life. And why the Great Reduction between the natural sciences feels far away. We can see quantum computation in some biological processes, as we saw with plants and robins. Perhaps it is not too much of a leap to suggest that we could use quantum to explain consciousness. But herein lies perhaps the biggest gap to close: the matter of determinism and randomness.

Flying is a true joy to me. Being so far above earth, where the minutiae of human life plays out in all its chaos and complication, is a breath of fresh air (although that is the last thing you get on a plane). Yes, I must admit, it is the sense of control. Looking down, getting a handle on things — physical time out. Let us look into that matter of control, because it is a big one. It unites many humans and human endeavours, but perhaps it also makes us blind to what Wallace calls 'higher considerations'.

The conquistador Francisco de Orellana was captured by the Spanish army and imprisoned in a small cold cell. He was soon summoned by a judge who sentenced him to death. To exert further mental torture on Orellana, who had led one of the biggest rebellions against the Spanish crown, the judge told him that he would not know exactly when he would face his executioner. The day when sentence was passed was a Sunday and the judge ordered that Orellana be hanged by the end of the week, and that he be told of his fate only on the morning of his execution. Imagine the horror of trying to fall asleep not knowing whether this would be your last night.

But hang on a minute, thought Orellana. Following the judge's logic, he worked out that the execution could not take place on the Friday because then he would know it a day ahead, on Thursday, and this would no longer inflict the surprise the judge had intended. Following his line of thought, he concluded that Thursday, Wednesday, and all the other days leading back to the Sunday when sentence was passed, were equally unviable appointments with the executioner, for the same reason that this would scupper the cruel plans of the judge.

Orellana had it all figured out on Sunday, and thought to himself that he had managed to turn his torment-by-uncertainty into something deterministic: the deduction that the execution could never come as a surprise!

This well-known riddle, often called the 'unexpected hanging paradox', has long been retold in many different ways that replace the execution with other ominous events, such as surprise exams, unwanted visitors, or letters relaying bad news. But to this day, philosophers argue about where the catch lies. In reality, it would be enough to keep the prisoner ignorant of the exact day, week, or even year, of the execution to keep them guessing to the end; the Japanese death row system works like this.

But not all uncertainty is bad. As a free man, I am much happier not knowing when I will die (it is probably something like tempting fate to write these words at 40,000 feet in the air). Similarly, Orellana would have been sustained by the uncertainty in his captors' ability to stop him escaping, no matter how small the odds — after all, people have busted out of Alcatraz — or that an earthquake might destroy the prison and set him free, or even that the judge might change his mind. All these things work to a prisoner's advantage and give him reasons to hope. And these are just the known unknowns. He might also be liberated by something he could not have anticipated, an unknown unknown.

Must we always live with uncertainty about the future? Can risks be minimised? Better still, could it be that if we really tried hard, we could eliminate all uncertainty, including those unknown unknowns? What has science to say about uncertainty?

INDETERMINISM

Whether uncertainty is a fundamental physical property or not has been asked for thousands of years, since the Ancient Greek philosophers. It was impossible for them to settle the question because their understanding of natural laws was vague. Medieval scholars found the issue equally difficult, and few pinned down the problem as eloquently and clearly as the Persian poet and astronomer Omar Khayyam (1048–1122) who summed-up the concept of determinism as the opposite of uncertainty in this magical verse:

'With Earth's first Clay They did the Last Man's knead,
And then of the Last Harvest sow'd the Seed:
Yea, the First Morning of Creation wrote
What the Last Dawn of Reckoning shall read.'

That is determinism in a nutshell: the very first moment determines all the other ones up to the last.

The most convincing illustration of this principle came with Newton's discovery of the laws of motion, which allow us to predict exactly the position of planets far into the future. That seemed to settle the debate opposing determinism and uncertainty. The notion that uncertainty might be a fundamental physical property was further banished when a French physicist called Pierre Simone de Laplace (1749–1827) devised an ingenious thought-experiment invoking an intelligent demon (yes, another demon). I will let Laplace speak: 'We ought then to regard the present state of the universe as the effect of its antecedent state and the cause of the state that is to follow. An intelligence, who for a given instant should be acquainted with all the forces by which nature is animated and with the several positions of the entities composing it, if further his intellect were vast enough to submit those data to analysis, would include in one and the same formula the movements of the largest bodies in the universe and those of the lightest atom. Nothing would be uncertain to him; the future as well as the past would be present to his eyes.'

So Laplace's worldview took hold, and it would be another 200 years before revolutionary discoveries in physics forever changed our picture of the world into something less deterministic and less mechanical. The spark that set fire to the dry old edifice came from Max Planck, whom we met earlier — the German physicist who realised that the classical physics of Newton and his later peers was incomplete. He is the one who introduced the idea of quanta, namely that energy always comes in tiny but discrete chunks. The far-reaching implications of this discovery cannot be overstated.

Yet not everything changed. The ability to make conjectures about the world and to settle them through experiments (and experiments only) meant that the scientific method survived and was strengthened during the quantum revolution that followed. It had proved its worth. Planck was clear that no conjecture was above the scientific method, including the age-old question about determinism versus uncertainty: 'from the outset we can be quite clear about one very important fact, namely, that the validity of the law of causation for the world of reality is a question that cannot be decided on the grounds of abstract reasoning'.

And this was taken seriously by Heisenberg, who demonstrated indeterminism with another thought-experiment, now known as Heisenberg's uncertainty principle. Imagine that, like Laplace's demon, we want to know the position and velocity of a single electron. To see it, we need to bounce light off the electron and then to detect the reflected light with a microscope. But here is the problem: light interacting with the electron will give it a push, so the position we see in the microscope will no longer be accurate. To make matters worse, the push will impart a random velocity to the electron, so we are uncertain about both the position and velocity. Heisenberg argued we can never know the position and velocity at the same time, no matter how sensitive our tools.

But there are two other ways — both stemming from Heisenberg's uncertainty principle — in which the quantum world is uncertain. The first is this. Suppose that you prepare a number of electrons (or atoms) in exactly the same physical state. This means that all measurements on each of them would lead to exactly the same statistical behaviour. Now, imagine that we randomly choose half of these electrons and measure their exact position, while we measure the velocity of the other half. What is interesting here is that the more precisely we measure the position on the first half, the less precisely we know the momentum of the second half. Heisenberg's uncertainty again — but, the bizarre thing is that there is now no interaction between the two halves! They were originally prepared independently and were kept apart during the measurement. Quantum uncertainty is spookier than Heisenberg imagined in his thought experiments. But things get spookier still.

The second scenario really bugged Einstein. He thought that it could bring quantum physics into conflict with his own cherished principles of relativity. Imagine a simple two-state quantum system, such as a spinning electron. Electrons are like small spinning-tops that spin clockwise or counter clockwise and point in any direction, horizontally, vertically, at 45 degrees, and so on. Astonishingly, if we carry out two consecutive measurements on the electron spin, the correlations between these measurements can exceed any correlations allowed by classical physics.

Spins measured at different times are allowed by classical physics to be correlated in the horizontal or vertical direction, so that if the first measurement yields say a 'horizontal clockwise' spin, then so does the second measurement. However, spin measurements on real electrons can be correlated in the vertical direction at the same time as in the horizontal direction (and all other directions!). This is because quantum mechanics allows electrons to spin simultaneously in the clockwise and counter clockwise directions — something that no spinning-top can do.

It also follows that the spins of two electrons can be more correlated than classical physics allows. And this bothered Einstein. It seems then that there is an excess of correlation between quantum systems and that this is independent of how far they are from each other. Worse still, it looks as though this excess correlation is transmitted instantaneously between the two, and so violates Einstein's axiom that nothing travels faster than light. An effect he derided as 'spooky action at a distance'.

The effect is not just a neat mathematical trick but has been observed and verified in laboratories many times. Such quantum correlations between objects and events are called 'entanglement'. Quantum physics stipulates that all the objects in the universe that have interacted — and pretty much all matter interacts via light and other fields — have become entangled. From this emerges a fascinating picture of our universe.

There is a general feeling of ambivalence about the notion of determinism. On the one hand, we feel that we have a certain degree of 'freedom of will'. We feel that our actions, at least some of them, are products of deliberate decisions we make after considering different options. Psychologically, many people are repelled by the possibility that all our actions are fully determined by our history and the environment.

However, although psychologically repelling, intellectually, determinism might be very pleasing. It means that everything that happens, happens for a reason. The German philosopher and mathematician Gottfried Leibniz called this 'the principle of sufficient reason'. If quantum randomness is genuinely inexplicable in terms of any other causes, this would clearly violate the principle of sufficient reason. There would be events in this universe (basically all quantum events) that would have no reason for happening the way they do. In other words, in the quantum world, things could always be otherwise.

Many people reject this view as mystical. Randomness always invokes the question 'why did it not happen some other way?' to which the only possible answer is 'because that's the way it is'. This answer is far from satisfactory especially in view of our rational side. In fact, the Dutch philosopher Baruch Spinoza, who was a rationalist par excellence, used the principle of sufficient reason to argue for a completely deterministic universe in which God is basically the sum total of all the causes. Here, the universe can only be constructed one way and even God has no choice in this — even God is bound by the fully deterministic logic.

It has long seemed that randomness and determinism cannot both be true within the same picture of reality. Here, however, quantum physics brings us an interesting alternative that is capable of embracing both randomness and determinism. Quantum physics presents us with determinism at the level of the whole universe, while still implying that any subsystem in the universe is fundamentally random!

If quantum physics applies to the universe (and this is, of course, yet another big if), the universe is then in a gigantic entangled state that evolves completely deterministically; yet, at the level of individual objects or chunks of the universe, there survives a fundamental uncertainty. The entanglement of an object with another creates that uncertainty. By that I do not mean the kind of classical uncertainty due to incomplete knowledge, such as whether the government will bail out a bank, but the quantum uncertainty described by Heisenberg arising from the physical interaction of matter at the fundamental level. There seems to be no escape from the uncertainty within the universe, since, crudely speaking, we would have to be outside the universe to perceive it as being deterministic.

And entanglement is the only route to uncertainty in the quantum universe. This means that any probability that we perceive as being classical, such as the probability that a tossed coin will land on heads, is in fact, fundamentally due to quantum entanglement. So is the randomness we encounter when trying to predict weather or the stock market, all apparent classical randomness is basically due to quantum entanglements with the environment.

One bonus that we get from this picture is the entropy of the universe, which quantifies the overall disorder, can be zero (since the universe in this picture is in a perfectly deterministic state), while within the universe we can still perceive the entropy of sub-regions as increasing (in accordance with the second law of thermodynamics).

Interestingly, this fits the emerging picture of the universe that, for lack of a better phrase, 'requires nothing to exist'. Recent measurements suggest that the universe has overall zero energy, charge and angular momentum. Zero energy just means that all the energy due to matter is cancelled by the equal (but opposite in sign) gravitational energy. Locally, we can have non-zero energy, but overall, there is none in this picture. Zero charge means that there is the same amount of positive and negative charge and we cannot create one without creating the other. Locally, we can have some negative charge, but this has to be countered by the same amount of positive charge elsewhere. Zero angular momentum means that the universe has no net spin. We can rotate ourselves, but the earth has to spin in the opposite direction (very slightly, since the earth is much more massive than us) to make up for zero total spin. The same now seems to be possible of entropy.

This picture would please Greek philosopher Epicurus. He claimed that 'nothing begets nothing', namely that something cannot be created out of nothing (because, as he says, then anything could be created out of anything). This view is contrary to the official Vatican position regarding cosmogony, but it is fully consistent with the modern physics. Locally we seem to have lots of stuff, such as matter, charges, motion, entropy, uncertainty, but globally none of these things exist, never will and never have existed.

In quantum physics, novelty arises naturally — for free as it were.

Could this have implications for how we understand the origin of life? Is indeterminism necessary for life?

How inanimate matter comes to life remains one of the deepest mysteries in science. If it remains a mystery, then the Great Reduction cannot take place — all natural science cannot be united. Perhaps this is our answer. The second premise of the 'survival of the fittest', the term coined by Herbert Spencer after reading Darwin's theory, is that the DNA containing the biological blueprint for building a living being undergoes unpredictable mutations when it is reproduced. There must be some uncertainty in that replication. Could it be that this uncertainty needs quantum mechanics to be genuinely uncertain? Maybe the second aspect of evolution would not even be possible if it was not for quantum physics.

The uncertainty about the final state into which the electron spin will evolve from an initial superposition is a genuine random phenomenon, and the best one can calculate using the full machinery of quantum physics are just the likelihoods of things happening, just probabilities.

This randomness is a fundamental unknown that no amount of knowledge about a system, no matter how detailed, can lift to predict with certainty which final state the system will evolve into. We can reasonably foresee a similar scenario taking place during genetic mutation. The chemistry of DNA replication involves exchanges of electrons and atoms, which are quantum objects subject to quantum superpositions, which evolve into physical states giving rise to new chemical bonds. While we might one day be able to predict why certain bonds are more likely to form, quantum physics means that we will never be sure that they will definitely form, so ensuring some randomness in the outcomes of genetic mutations.

The uncertainty in chemical bonds, or any other uncertainty, as we have seen, can be thought of as the entanglement between atoms forming these bonds and their environment. It is because of the highly interactive nature of physical processes that things become entangled with the important consequence that all parts become fundamentally random.

To see why uncertainty is important here, we can imagine that random mutations are like different gambling strategies, each with a different likelihood of success. Evolution is then like a gambling game between an individual and its environment. Profiting in evolutionary terms means that, regardless of the environment, life will propagate and losing means that life will end. So, the game is as follows: the individual produces copies of itself, but, importantly, the copy will be slightly different due to random mutations.

These mutations alter slightly the set of properties of the new individuals, which are then tested in the environment. The number of individuals produced depends on the gambling strategy. Any new individual will have to be fed and nurtured and, given finite resources, reproduced in as many copies as resources allow.

The new individuals then either pass the evolution test and multiply or they fail and die. It is now clear that those who survive profit more and more because their strategy fits the environment better and better, albeit at the expense of all those who did not make it.

But there is a catch: the more profitable that life becomes, the less profitable its environment. And as environmental conditions change as resources dwindle, life finds it harder to propagate and might not be able to adapt fast enough to the changing environment. And so, the arms race between the individual and its environment continues, the biological boom and bust from which springs all the wonderful bio-diversity surrounding us.

And without those genuine surprises that can be explained quantumly, one of the key resources driving the evolutionary cycle of boom and bust that propelled humans into existence would be lost. Nasty surprises are equally plausible, and, like Orellana, we could find out that our luck ran out when the human race became too successful for its own good. Or back to Wallace's words, that our pursuit of wealth leaves us 'blind to higher considerations'. Perhaps randomness is a crucial strategy in evolution?

However, while randomness might matter for living systems, it is something that is clearly important for inanimate objects. And what this leads to, of course, is the question: what is the difference between life and no life? How can *that* gap be explained?

LIFE AND NO LIFE

I'm going to invite a few wise thinkers to join me on this flight, to share their thoughts on this subject. The first is Addy Pross. This Israeli chemist firmly believes that the line separating living and non-living molecules is not actually that sharp. He says: 'We now know that a mechanism akin to

Darwinian evolution actually operates, in the first place, on non-living matter — even on single molecules. Feed a population of RNA molecules the appropriate chemical building blocks and, under the right conditions, they will start to self-replicate. What's more, over time, you will see the population evolve: slow replicators will give way to faster ones'. RNA is not living material in any meaningful sense, yet it is subject to evolution. Thus, we find our first bridge between living and dead matter.

His second statement is even more momentous. Evolution exhibits an identifiable driving force, a direction if you like, and this 'teleological' tendency acts at both the chemical and biological stages; that is, it operates both *during*, as well as after, what we think of as abiogenesis. Thus, the purpose-driven character of life, the very thing that seemed to distinguish biology from the rest of nature, turns out not to be unique to life after all. Its beginnings are already discernible in certain inanimate systems, provided they are replicative and able to evolve. And this driving force can be described in strictly physical terms. Put simply, it is nature's drive towards greater stability — a drive that is as ubiquitous in physics as it is in biology.

A very different school of thought comes from the grandfather of quantum physics, Niels Bohr. In his famous lecture in 1932, 'Light and Life', he argued that we cannot, even in principle, probe life *in vivo* to understand it. In Bohr's words, 'the existence of life must be considered as an elementary fact that cannot be explained'. Just as Planck's constant — 'which appears as an irrational element from the point of view of classical mechanical physics' — nevertheless forms an irreducible foundation of atomic theory, so too must life be taken as an inexplicable starting point in biology. Bohr's opinion, however, is a notably pessimistic one, and others have had higher hopes for the unification of scientific realms.

Schrödinger also emphasised the idea that life tries to maximise *free energy*, namely the energy available to do useful work. This is another way of saying that it wants to stay away from equilibrium — unlike a stone for example, which, when left to its own devices, just stays as it is and does not try to do anything useful.

Here is a little truism for you. *Unchanging things do not change, and changing things do change — until they change into things that do not.* That statement is, of course, true as a matter of logic: as true as 'one plus one equals two'. And like 'one plus one equals two', it turns out to make surprisingly strong predictions about how the world works. If every changing thing really can change into an unchanging thing, then we should expect *all* changing things to change into unchanging things eventually.

The separation between sciences is crumbling. Nature does not recognise disciplinary borders, and as we deepen our understanding, we see more of what these traditionally distinct branches of science have in common. There remain, however, curious hold-outs.

Physics deals with the basic properties of matter and energy and how they interact. Chemistry is concerned with how atoms get together to form more complex molecules and the effect this has on the resulting substances. What both have in common is that they study inanimate matter.

Biology, on the other hand, studies living organisms. And here, we encounter the central obstacle to seeing all natural sciences as one big coherent whole. Inanimate matter seems to obey the strict laws of nature without exception and down to the last letter. Living things, by contrast, appear to have a will of their own. They are best understood — perhaps even best defined — by what might be called purposiveness. They have an agenda, and while they cannot violate the laws of nature, they certainly can exploit them to their greatest advantage in order to realise their goals. Inanimate matter does not do that.

I am not merely asking whether living systems can exploit the stranger aspects of quantum physics to improve their chances of survival. The simple answer to that is, yes, it appears they do. There is evidence to suggest that even the quirkiest of quantum effects, quantum entanglement, is used by photosynthesising plants to channel light energy towards their energy-producing parts by the most efficient route. Similarly, some birds are thought to use odd quantum effects to detect the earth's magnetic field during migration. The efficiency advantage that quantum physics could be giving these living systems is that it allows them to perform several tasks at the same time, something computer scientists call parallel information processing. Very few people expected that the full repertoire of quantum physics could survive in macroscopic, warm and wet, noisy environments like plants and birds.

But this has little to do with reducing biology to physics. Life also exploits classical mechanics and gravity, and that does not mean that classical mechanics and gravity can explain the evolution of life itself. Life could be consistent with all the laws of physics, and we still might require principles in addition to physics to explain it. In fact, most biologists would agree that it is indeed consistent with the laws of physics, in the sense that it must obey all of them. It not only exploits physics, but is clearly also affected by it: clearly enough, the environment affects living beings via physics. What we want to know is whether we can go beyond questions of 'exploiting' and 'being affected by' physics to consider the possibility of a more profound relationship.

The question here is whether evolution, which is one of the pillars of the science of biology, is fully a consequence of physics. In particular, is it a consequence of quantum physics, which explains everything we know about atoms and molecules?

Back to Addy Pross, who fronts the latest chapter in the attempt to derive biology from physics. He suggests that, very much as inanimate matter conforms to thermodynamics by maximising entropy, living beings strive to maximise what he calls 'kinetic stability'. This is not the same as maximising entropy production. Rather than reaching a passive state of static equilibrium, as all inanimate matter invariably does, according to the second law, living systems achieve a dynamically stable state, but they have to keep working in order to maintain it. The dynamically stable state is fragile and needs constant re-establishment. Picture a bird flapping its wings only to stay suspended in one place in the air: this requires careful balancing, and though it is clearly dynamic, it still results in a stationary condition.

If Pross is right, we might have the ingredients to reduce the key features of evolutionary biology to chemistry. And given that chemistry is reducible to quantum physics, it seems as though we might be able to go all the way from biology to quantum physics. This would be a great achievement. However, like any other great achievement, it raises questions.

We started by saying that what discriminates living from non-living systems is a sense of purpose. If biology is reducible to quantum physics, and typical quantum objects such as atoms and molecules show no sense of purpose, where does the transition occur? Where does the 'desire' to achieve the state of kinetic stability come from? This, of course, brings us back to square one. One easy way out is to conclude that purposefulness is simply an illusion. Pross would probably say that it is an emergent property that arises when chemistry becomes complicated enough. But given that this sense of purposefulness is how we identify life in the first place, perhaps we should resist conclusions that seem to wave it away too easily.

The most interesting recent development on this issue is by a physicist called Jeremy England. He took the idea of Boltzmann and Schrödinger that living systems are sophisticated thermal engines, what Monod called 'living Maxwell's demons', that are much better than inanimate matter at taking useful energy from the environment and converting it into useful work. The price to pay is the heat generated in the environment.

What England did is try to use thermodynamics to write down the mathematical laws of biological evolution. This is the closest I have seen anyone trying to make a physical theory of evolution, which is something I believe would make biology closer to physics and physical sciences such as chemistry. How would this approach enlighten us as to why living systems appeared in the first place out of an initially dead universe?

The answer would be that life is a thermodynamical necessity. Once the laws of thermodynamics run their course on inanimate matter, they naturally create the pressure for living systems to evolve as more optimal energy users. England is the latest word in the story that started a long time with Boltzmann. If thermodynamics is essential to life, this might make us take the thermodynamical view of gravity more seriously, and as a consequence, view the existing quantum gravity gap in a new light.

I do not want to appear to pretend to have the final answers to any of these questions. It is easy to fly over the earth and fit everything into your neat picture. And yet, the speed of progress in fields that cross the boundaries between natural sciences, not least quantum biology, makes me optimistic that we will get some sooner or later. At any rate, I would not bet against it. For now, however, we will just have to keep digging deeper and deeper.

Well, we have just had our 'crew ready for landing' warning, so I had better wrap up. We have journeyed through a few rather sizeable gaps in natural sciences. We have seen that the gaps could be fundamental if the only way to bridge them is using computer simulations on universal

Turing machines. But, there are many connections still to be made before we might encounter the Halting barrier — if ever. An optimistic scientist, and we all are in this respect, will keep working on bridging the gaps between physics, chemistry and biology.

We still have not heard from a few important dinner guests back in Oxford. The economist with the booming voice, despairing over the latest government requests to predict the next stock market crash. Or the Japanese social scientist, trying to explain the actions of humanity in groups and countries. These are the topics of the day, vital if we want to understand the next chapters of our future. And even the most optimistic researchers might feel that the next gap — this is between the natural and social sciences — is beyond the reach of science. Can we understand human actions, financial transactions and social trends, the same way we understand atoms, molecules and biological cells?

To bridge the two halves of the book — natural sciences and social sciences — I would like to share a conversation I dreamt up the other day. The three characters span six different fields between them — quantum physics, computer science, biology, philosophy, politics and economics. To guide us through this transition, meet the following:

- **David Babbage** is a theoretical physicist, who argues for the quantum perspective in computation, but is not sure about biology.
- **Margaret Ricardo** is a sociologist, and largely neutral as far as quantum effects in biology are concerned.
- **Simplicio Smith** is an experimental physicist and a dedicated follower of classical physics.

Margaret: I have been reading a lot about quantum biology recently — there seems to be quite a bit of controversy regarding the presence of quantum effects in biological processes.

Simplicio: I do not think there ought to be any controversy, really. Biological systems are large, warm and wet, and are, therefore, classical.

Margaret: Hmmm, I do wonder. I have read quite a bit in the popular press about high T superconductivity. Ok, granted, these superconductors are not wet in the usual sense of that word, but aren't they actually large objects that behave quantumly at temperatures of about 150 Kelvin? I remember one of the researchers saying that all electrons in high T

superconductors are impervious to noise because they are entangled —
I think he said that they lived in a 'quantum protectorate'. Nice phrase. But
David, you must know much more about this kind of stuff?

David: Well, I have really thought more about the role of quantum
mechanics in deliberately designed human information processing. I
have not been following much about any naturally evolved information
processing.

Margaret: Ah, quantum computers! I have read about them too. It
is all very exciting.

David: Yes, they will be great. And it is clear to us that quantum
coherence is very important to achieve the full potential of quantum
computers.

Simplicio: I thought that the most important quantum computations,
Grover's search and Shor's factorisation, could in fact be done with
classical waves — there is no need for quantum physics. After all, the
term coherence you mentioned earlier, originates in the classical theory
of waves. I remember this experiment where some group performed the
search algorithm using only a diffraction grating and classical light.

David: Classical light? I have a problem with that phrase, but let us
not go into that now. You can indeed execute computations classically, but
you will always pay a price for not being quantum. In the example you
gave, the number of slits in the grating is equal to the number of database
elements. If, instead, you use quantum bits, you only need a logarithm
of that to store the database. This is why quantum computers are more
efficient.

Margaret: I thought you said it was to do with quantum coherence.
Are you saying that encoding things into quantum bits is crucial?

Simplicio: That cannot be the case, since I can encode the database
in classical bits and that would still be equally better than the diffraction
grating.

David: Hold your horses. I think you are starting to mix things
up. When it comes to efficiency, we have to be careful to talk about the
resources we are trying to optimise. One obvious one is time. We want
to compute as fast as possible. Another one is space, or, in other words,
the size of computer memory. Then, we might want to optimise the total
energy expenditure or precision and possibly many other things. Encoding

with classical bits would indeed save on the memory requirements, but you would be very inefficient in terms of time. Somehow, whatever —

Simplicio: I am sorry to disagree with you again, David, but search has been done efficiently with a single atom encoding the database, so it cannot be due to entanglement. You know this well, yourself.

David: I was really hoping not to have to go too much into technicalities. After all, Margaret is not a physicist and I do not want to bore her with all the —

Margaret: Oh, come on! Do not hold back because of me. I may not be a scientist, but I read about it all the time. I call myself a closet scientist, or a closet quantum mechanic. Ha.

David: Ok — you asked for it! Well, it all depends on how you define entanglement. If you require objects to be well separated for your definition of entanglement, then an atom would not qualify. But if you only need to be able to define two different subsystems, then an atom is fine. I would say, to be on the safe side, we need quantum coherence. This is why I said it was quantum coherence that was crucial.

Simplicio: Indeed, but this is precisely what I was objecting to. I do not see that you cannot do computations efficiently with classical waves only.

David: It is true that we do not have a strict proof of that. I am not even sure if a strict proof is possible. However, going back to your atom as a database, there is another inefficiency associated with it. It is not discussed much in the literature, I do not know why. If each level corresponds to a database element, we again have a huge problem with the inefficiency of encoding the database (just like in the case of the classical grating).

Simplicio: Let me stop you right there. The levels in the atom need not be (in fact, in reality, they are not) equally spaced. So, we can cram as many as we like into a small space. Surely that is efficient in terms of memory.

David: I'm afraid not. If any levels are closely spaced, then we have a problem of resolving them in our experiments. So, we might be efficient in terms of space, but the read-out gets harder. Namely, the cost in terms of precision has got to blow up.

Margaret: I am beginning to like this. It is starting to sound like a Catch-22.

David: Ha. I would call it sod's law. It seems to me that there is always a trade-off between resources. If we optimise on one, another resource becomes more demanding. The only way to be fully efficient is to have encoding in terms of qubits and allow unlimited quantum coherence between them. Like I said, I cannot prove this formally. Yet.

Simplicio: I rest my case. Anyway, earlier you said you had a problem with classical light. What is the deal with this?

David: The deal as you say is that there is no such thing as classical light. All light is really quantum. Classical light is just an approximation when we talk about a large number of photons present in the beam of light. Which brings me to another point I was going to make. A large number of photons. If you think of optimising the number of photons (and why not? After all it is the same as optimising energy), then you do not want to be using classical light.

Simplicio: Classical light is a lot easier to produce. It is all around us. Producing single photons is damn hard. Irrespective of that, I still think that classical physics should be able to match quantum computation and not lead to an abuse of resources. To me, it indicates that classical physics is enough — discord is closer to classical correlations than entanglement.

David: Closer how? In terms of what?

Simplicio: It does not violate Bell's inequalities. It has no non-locality. No 'spooky action' is needed to account for discord.

Margaret: Discord is not spooky? Did I understand that right?

David: You did. And there is no contradiction here. Expressing it in your, or actually Einstein's language, spookiness has nothing to do with computational efficiency. Anyway, Simplicio, I do not really like the term non-locality since quantum mechanics is a local theory. It can be made to comply with special relativity and there is no faster than light communication even with entanglement.

Margaret: Really? So why was Einstein worried?

David: He was worried because the implication is that the world contains genuinely random elements and he disliked randomness as much as non-locality. Entanglement forces us to either give up locality or to give up reality (since quantum indeterminism means things do not exist until you measure them). Einstein felt that with quantum physics, he was stuck between a rock and a hard place.

Simplicio: Yes, ok, David. Margaret, I know you read a lot of popular science, but this discussion seems to be taking us well away from the starting point. I made a comment that biological systems are classical because they are large, warm and wet. Irrespective of quantum computing, and certainly irrespective of non-locality, I still stand by this point firmly. I have been doing a lot of modelling of biological processes recently and have found complete agreement with experiments without any need for quantum physics.

Margaret: Really? I thought I read that the periodic table of elements cannot be understood without quantum physics. Since biology is underlined by chemical processes and these, as far as I know, involve interactions of atoms, I would expect biology to be very dependent on quantum physics.

David laughs.

Simplicio: Well, obviously. But this is what I would call a 'trivial' role of quantum physics. We all know that the stability of matter relies on quantum physics. Atoms and molecules cannot exist in classical physics. They would literally collapse. Take this for granted and you do not need any other fancy effects such as quantum coherence, entanglement and so on.

Margaret: Funny. That is not what I read a couple of days ago in the Daily Mail. An article was claiming that a bunch of boffins from Cambridge and Hong Kong, or whatever, I do not remember, found that birds, European robins it was I believe, used quantum entanglement to migrate to Africa. I remember being shocked and excited by this.

Simplicio: Yeah, right. They sound like a bunch of theoretical physicists and you know that they are capable of proving anything, especially things that are not true.

David: Ahem. I am still here.

Margaret: No, no, I distinctly remember there were some experiments mentioned —

Simplicio: Well, I am sure it is all circumstantial evidence. I imagine these experiments have a large error and are not very conclusive. Anyway, what has that got to do with entanglement?

Margaret: Oh. I cannot remember. David?

David: Neither can I. But if you allow me to speculate...

Simplicio: Here we go. Now we will see the kind of stuff theoreticians are capable of conjuring up.

David: I will pretend I did not hear that. Well, I am familiar with chemistry experiments where the outcome of the reaction is influenced by changing the nature of entanglement between two electrons in the experiment. They could be in two different states and the two states lead to different chemical products. The crux of the experiment is to apply an external magnetic field to alter the ratio of different states and hence, affect the output product of the chemical reaction under control.

Simplicio: Let me stop you right there. Surely you will not be suggesting that the electron entanglement is crucial here. This is precisely what I would call a trivial quantum effect. After all, electrons are in a single state because they presumably occupy the same state initially. It is all due to the Pauli Exclusion Principle, much like the structure of atoms we were discussing here.

Margaret: What, according to you Simplicio, would then be a non-trivial effect?

David: Good point, Margaret. What indeed?

Simplicio: Non-locality, for instance. It has got to be spooky!

David: But we have already talked about the fact that non-locality has nothing to do with the efficiency of quantum computation. Why should it be any different in natural information processing? Speaking of efficiency, by the way, it is not clear at all that the resources nature is trying to optimise are the same as the ones we are trying to optimise when building computers.

Margaret: The resources you were talking about before are very general, on the other hand. Space, time, energy, precision. Should they not matter even more in the natural world? After all, ever since Darwin, biologists keep emphasising the fact that nature has no foresight. It seems that natural processes work under even stricter constraints since, unlike us, they cannot see too far into the future.

Simplicio: Good point. The reason why we do not see a fully-fledged quantum computer evolved by natural selection is that it would simply cost too much effort (compared to the benefit it might present) to produce it. Even more importantly, mutations have to be available to push in the right direction and the external conditions too. I just do not see —

David: Hold onto that thought. I want to come back to it. But first I would like to address the point I was making about resources. Ok, maybe the resources in the living systems are similar to computational, but surely there are important differences. For instance, in quantum computation we are always minimising the time taken to execute something, while keeping other resources tractable. But in biology, I can well imagine that sometimes we want to slow some things down.

Margaret: Rate of cancer growth?

David: I am serious. Biological processes are complicated, they might require careful timing so that one process does not start before other processes have finished. Someone was telling me the other day that there is an electron transport process in the mitochondria. It is not well understood, but the point was apparently to get an electron from place A to another place B in the mitochondria. There is chemistry going on at both ends A and B. These chemical reactions have their time scales and it does not pay for the electron to arrive at B faster than the chemistry can afford to accommodate it.

Margaret: But this is not all that different to parallel computation. One processor has to wait for the other one to finish in order to take over. I thought this is how all computers work, in fact.

Simplicio: Yes, bits of computers need to be synchronised, but this is nothing to do with the classicality really. The point is that the electron hopping from A to B is a classical process. I have modelled this many, many times before. I always use the classical diffusion equation and my theory is in full agreement with experiments.

David: Ok, but where do the rates for hopping come from? Surely you use quantum mechanics for that.

Simplicio: Indeed, but this is what I called 'trivially quantum'. The point is that there is no spatial coherence in electron transport. You would call this classical, viewing it from your computational perspective.

David: So, Fermi's Golden Rule is trivially quantum? It seems to me that you call everything trivially quantum in biology and discount it in this way. I do not see what you would count as a genuine quantum effect in biomolecules. The fact that a lot of biology can be modelled classically does not mean much. It simply means that we have not probed the right domain. Surely, classical physics will fail in some places. I can go back to light and assume processes that depend on single photons. They have got to be genuinely quantum. There simply is not any classical analogue of a single photon.

Simplicio: Yes, there are places in biology where we have single photons. Are they relevant? I mean, do they have a biologically functional value? If instead of single photons, we used a coherent superposition of them we might — I suspect, we will — get the same answer.

David: In other words, you are saying that single photons are trivially quantum. Not again!

Simplicio: I will change the tune then. You have talked earlier about quantum computers. They have to be carefully engineered by us. Qubits need to be encoded into stable physical systems (usually at low temperature to minimise the noise) and then, they have to be coherently driven to perform quantum computations. I do not think that nature drives or prepares living systems in the same way.

David: Are you asking if nature has evolved quantum physics? I do not think this matters. If bio processes are quantum, and surely, they ought to be, then that is it. Even the fact that similar things could be done classically is irrelevant. There is no choice. If the fundamental laws are quantum, that is what nature has to work with.

Simplicio: No, what I mean can be illustrated using the energy transfer we spoke about earlier. For energy transfer to be quantum, the light driving it also needs to be quantum coherent. But natural light simply is not coherent. To get coherent light, you need to build a laser. There are no lasers in nature as far as I know, arising spontaneously.

David: I am still not sure if this argument is relevant. If we have single photons in bio systems and they propagate, then whatever follows most likely needs the full quantum mechanics to explain it. I mean you need to evolve the state with the Schrödinger equation. Even you surely would admit that this is non-trivially quantum. It has simply got to be that way.

Margaret: I just finished Schrödinger's book 'What is life', actually. He does make a convincing claim that explaining life might require some other laws of physics. It is usually interpreted that by other laws he meant quantum. That would support what you are saying, David.

Simplicio: But you are forgetting the chapter 'Why are atoms so small'.

Margaret: You are right — I do not remember that one at all...why are they so small?

Simplicio: Schrödinger says that the question is not asked the right way. We should ask instead 'why are we so big?'.

Margaret: Is that not just the same question asked differently?

Simplicio: No. Once you assume that atoms are the building blocks and that they behave randomly — that God does play dice kind of randomness — then you realise that they are very unreliable when in small numbers. I mean, they just do all sorts of unpredictable things. However, when you combine many atoms together, the quantum noise gets washed out and we are left with a more or less deterministic reliable classical device. So, this is why living systems are large compared to atoms. Life needs accurate, predictable processes, such as cellular metabolism, division and so on.

David: Ah ha! Finally, we are having a debate. I like this much better than your 'biology is classical, because all quantum stuff in it is trivially quantum' argument.

Simplicio: Oh, c'mon. I have effectively been saying the same thing as Schrödinger.

David: Ha! Right... Sure. This gives me an interesting idea. It is a twist on what we have been talking about. Sounds a bit topsy-turvy, but the more I think about it, the more I like it.

Simplicio: Here we go, another top-tier journal paper in the making...

David: Hear me out. We have been asking if nature would bother to evolve quantum behaviour. But maybe we should follow Schrödinger and think that nature needs to evolve classical physics.

Margaret: Einstein would have liked this...would he not?

Simplicio: I very much doubt that. He wanted the kind of classical reality to be more fundamental than the quantum one. Go on, David.

David: Imagine that nature gets given (as it does) raw quantum elements, that are random and on top of it, randomly distributed (looking at some biomolecules, they really look random when compared to ordered physical systems such as periodic crystals). However, what you need is to get them to perform something deterministically. Say you would like the molecule to be able to transport an electron across, as we talked before. Then, you would not like to rely on quantum physics as to get coherent deterministic quantum behaviour requires a great deal of coordination and external control. However, if you somehow combine lots of quantum elements, ultimately, they might get classical and therefore more predictable and deterministic. So maybe classicality has 'deliberately' evolved from quantum physics as a form of error correction.

Margaret: I just thought of an interesting possibility. Small objects behave quantumly. The whole universe, cosmologists say, is also quantum. That would make life exist in a classical region right in-between two quantum domains. Would that not be strange?

Simplicio: That is interesting, Margaret. Maybe there is a sweet spot between the two quantum extremes which has perfect conditions for life. Life exists where complexity is maximal, and both extremes of the universe are quantum and therefore simple. Somewhere in-between is where complex things happen. But wait a minute — one aspect we completely ignored is heat. The main limitation to miniaturisation of computers is the fact that the resulting heating is so big that newly-designed chips keep blowing up. We, in fact, do not know how to cool the computer down fast enough and that is currently the main limiting factor.

Margaret: Really?

David: Yes. And the answer to this is 'reversible computing'. We actually know that we can make all computational steps fully thermodynamically reversible so no heat would be generated.

Margaret: Quantum computers again?

Simplicio: No — no need for quantum computers. Classical computers can be reversible just the same. Curiously, in fact, DNA replication, which you can think of as a kind of natural information processing, is much more efficient than our computers. DNA wastes 100 units of heat per computational step, whereas our computers use about 10000 units of heat. The forecast is that our technology will match DNA by 2020 and by 2030 will hopefully approach the reversible level.

David: That is very interesting. I mean, that we are still not as efficient as the DNA.

Margaret: Well, nature did have a 4-billion-year head-start. But, of course, it has no foresight as they say. Evolution, unlike us, is blind. Natural selection seems the dumbest way to engineer anything. Mind you, some biological mechanism look mind-blowingly sophisticated. But, and I do not really want to change the topic from this exciting one, however, it strikes me that human societies — something I have learnt a lot about in my undergraduate days — might also have evolved naturally. Does that mean that human societies are also efficient thermodynamically speaking? I mean, why should not most of the physics we have been discussing apply to human societies?

Margaret clearly has a point. Why not? And it is an inspiring idea to lead us into the next gap.

CHAPTER V

ECONOMICS

If making predictions and testing them against experiments is the key to scientific progress, then economists and other social scientists are somewhat doomed. People are slightly more difficult to predict than atoms. As fund manager Peter Lynch said, 'There are 60,000 economists in the U.S., many of them employed full-time trying to forecast recessions and interest rates, and if they could do it successfully twice in a row, they would all be millionaires by now. As far as I know, most of them are still gainfully employed, which ought to tell us something.'

Economists just about have a handle on transactions between two people and how to predict them. But zoom out to the macro? That is where the trouble lies. 'There are two kinds of forecasters,' John Kenneth Galbraith, a twentieth century economist, told us. 'Those that don't know, and those that don't know that they don't know.' Tell me something I do not know, I hear you cry. We witnessed a huge financial collapse in 2008, which I daresay a large portion of our population felt in some way. It was not unprecedented, and it would not be the last, and yet still very few experts saw it coming. There are movies now made about those that did — this is how rare good predictions were before the collapse. Major financial crashes happen about once per generation. How can they be so complex to predict?

Well, because economics is, broadly speaking, the science of how humans make their choices. Which makes it perhaps the most complex science there could be, given our tendency towards irrational behaviour and unpredictability. If you remember Will Hutton's dinner party, and the economist I met with the booming voice. We spent a large part of the end of the night discussing the struggle of the economist. I find economics perhaps disproportionately fascinating. And in particular, economics' biggest gap: using an understanding of individual behaviour to then understand wider human behaviour. I believe if this gap could be bridged, so could all gaps in social sciences. This is a big one to hypothesise about. Anything involving human behaviour is unendingly complicated, and where better to contemplate it but here, in Dubai, at an annual meeting of the World Economic Forum.

I have the privilege of being part of the annual meeting of the Global Future Councils, one of 700 members who meet to discuss opportunities enabled by breakthrough technologies. Note the word *opportunities*, somewhat different to the topic of Hutton's Oxford dinner party: *disruptions*, the focus here being on how technology can be turned to our advantage. (Does this mean I should perhaps look at my 13-year-old son's incessant drumming as an opportunity rather than disruption? An opportunity to practice tolerance?) There is certainly an air of optimism, excitement, void almost entirely of the chuckling English cynicism of Hutton's dinner party. It is not just the WEF — it is Dubai as well. An electric optimism that cuts through the city's strange beginnings, a clean, air-conditioned beacon of what humans and technology can achieve when combined with the word *opportunities*.

I have come to quite like this strange place. It is not a million miles from Singapore, actually, which has become my second home over the years. It is the perfect setting for the meeting of the World Economic Forum, as well as this chapter — a place where hypothesis becomes reality.

How do individuals make their choices? How is this affected when another person comes into the equation? And how can we get from understanding micro behaviour on this small scale to understanding large group behaviour?

What happens if, during my few days in Dubai, Metallica happened to be playing in the city? Obviously, I go. But then I find out that the very night their gig is on, my old university friend is also in town for one night only, and free to meet for some drinks in the Burj Khalifa (highest bar in the world in the tallest building). Going to see Metallica and meeting my dear friend are two mutually exclusive events and there is no way I can do both. The Metallica ticket costs $100, and a night in the Burj Khalifa would be about the same. If I do not see my friend, then I would not see her for at least another year. How would an economist decide?

First, why an economist? Why not a psychologist, or a sociologist, or a priest? Well, because economics deals with how people make choices with scarce resources. If the resources were not scarce, in this case, time being scarce, the choices would be easy. The fact that in this scenario we are talking about a single individual (my humble self) makes this a question for microeconomics (macroeconomics would be concerned with how such a decision would be taken on a country-level).

The answer would very much follow the original Cartesian method of choosing between different alternatives. René Descartes, a seventeenth century French philosopher, argued that you should write all the pros and cons for the alternatives and give them different weights (positive for the pros and negative for the cons). Then sum them all up and the alternative with the highest score wins. That is an easy one for me to do. Yet to use this method for even a small number of people, say 10, and come up with a compromise, is almost impossible. All 10 people will invariably have different preferences — even this would not be a big problem as long as there is consistency in their choices. This oft-quoted restaurant exchange shows how an introduction of a new option can add more confusion to the previous two options.

Waiter: Good evening, madam. Can I get you a drink to start off with?
Madam: Yes, please. What kind of juices do you have?
Waiter: We have apple juice and orange juice.
Madam: OK — I will have the orange juice, thanks.
Waiter: Excellent choice, madam! Oh, I just remembered. . . we also have cranberry juice.
Madam: You also have cranberry? Well in that case, I will have the apple juice!

(Note: This would not be possible in the Burj Khalifa, where the minimum bar spend per person is an average of $90! A lot of fruit juice would have to be consumed — she would have to take all three for starters.)

Adding a new option should not change our initial preference between the original two. Yet, the human mind is not rational and consistent, and the problem of three adds even further complication. A slightly less often quoted exchange: My three children are off to the local store to buy some tuck. They each want different sweets, but they do not have enough money (as my last grant proposal was rejected) so they have to make do with one to share. Mikey prefers Haribo over Mars bar which he prefers over Nutella. Mia prefers Mars bar over Nutella and the last is Haribo. Leo, finally, prefers Nutella (he would live off the stuff if he could) over Haribo which he prefers over Mars bar.

This seems easy to settle. Haribo beats Mars bar by two to one and Mars bar beats Nutella by two to one. Therefore, you might think Haribo ought to beat Nutella. But, strangely, Nutella beats Haribo by two to one as well, making it impossible to choose the top preference consistently.

On a scarier scale, this kind of ambiguity features in some political voting systems where the candidate elected could end by not being favoured by the majority.

How do economists deal with these small-scale transactions? The quotes that opened this chapter from Galbraith and Lynch show us that even practitioners of economics are aware of the difficulties in making predictions, because of the complexity of macroeconomics. This is why economists are frequently distrusted and even disliked. They are unfortunate in how often they star in jokes because of the big gap between their theories and reality.

A man is walking along a road in the countryside when he spots a shepherd with an enormous flock of sheep. He challenges the shepherd: 'I will bet you £1000 against one of your sheep that I can tell you the exact number in your flock.' The shepherd thinks it over. It is a big flock, so he takes the bet. '973,' says the man. The shepherd is astonished, because that is exactly right. He says: 'OK, I'm a man of my word — take a sheep from my flock.' The man picks one up and begins to walk away.

'Wait!' cries the shepherd. 'Let me have a chance to get even. Double or nothing that I can guess your exact occupation.' The man thinks about it, then shrugs. 'Sure.' The shepherd says: 'You are an economist for a government think tank.' 'Astonishing!' exclaims the man. 'You are exactly right! But tell me, how did you deduce that?'

'Well,' says the shepherd, 'put down my dog and I will tell you.'

Is economics, and indeed any other social science, so complex that we are unable to apply the standard scientific method to their problems? Will our predictions always be so bad as to lead us to mistake a dog for a sheep?

Not necessarily.

Andrew Yao, the director of the Centre for Quantum Information at Tsinghua University, is a distinguished computer scientist who applied his knowledge to many fields, including quantum physics and economics.

I met him when he was reviewing research at the Centre for Quantum Technologies (CQT) in Singapore a couple of years ago. Based on this review, the Singapore government would decide to continue with the funding or pull the plug. The panel was to report to the government at the end of the review. The report to the government was extremely effective because, when asked what he thought about the centre, Yao said: 'The centre is great. If you don't want to keep funding them, I will buy it lock stock and barrel, move it to China and increase everyone's salaries by 25 percent!' That did the job. The CQT continues to thrive in Singapore.

The problem that established Yao's reputation and sits at the boundary of computer science and economics is known as 'Yao's millionaire's problem'. There are two millionaires and they would (for some reason) like to establish which of them is wealthier — but without revealing the exact extent of their wealth. This is actually an important problem in e-commerce and data mining. The former concerns all online financial transactions, while the latter is a subfield of computer science that looks for patterns in large, seemingly unstructured databases.

I will give you a flavour of how the millionaire problem can be tackled by introducing another version of the problem, known as the socialist millionaire's problem.

Two employees of Starbucks, Alice and Bob, would like to know if their wages are the same. They live in a society where people believe they ought to be paid the same irrespective of their contribution or effort (socialist). But even though they believe in the equality of income, and they want to know if they are paid the same, they are still rather proud and secretive about the respective amounts they receive. Can they somehow communicate with each other to establish if the income is the same without disclosing the amount?

The answer is yes, and this is the protocol. Imagine for simplicity that they make either £100, £200, £300 or £400 a month. Bob goes to an office supply store and buys four lockable suggestion boxes (with different matching keys). He labels the four boxes as £100, £200, £300, and £400. Bob discards all of the keys except the key for the £200 box (because that is how much he makes per month). Alice sees him keep only one key, but she does not know which box the remaining key belongs to.

Bob then gives the locked suggestion boxes to Alice. In private, Alice puts a slip of paper saying 'yes' into the £300 box (because that is how much she makes per month). She puts slips of paper saying 'no' into the other boxes.

Alice gives the boxes back to Bob who, in private, uses his key to unlock the £200 box and get the slip of paper inside. Bob sees that the slip of paper says 'no', meaning Alice does not make £200 per month like he does. He tells Alice that they do not make the same amount of money.

However... Bob now knows that Alice does not make £200, but has not learnt if she makes £100, £300, or £400 per month. Similarly, Alice now knows Bob does not make £300 per month, but has not learnt if he makes £100, £200, or £400 instead. Hmm. Naughty Starbucks.

Here, the protocol between Alice and Bob can be executed by two computers (hence, applications in e-commerce). Yao's millionaire's problem can be solved in a similar way. Unfortunately, not all problems in economics can be computed and solved in such a way. Things get much harder in other economical applications, and sometimes an optimum solution does not exist. Moving from a transaction such as this between two people to larger-scale predictions is where problems come. And on top of it, there is discord between what mathematics tells us the best solution is and what our emotions dictate.

Galbraith and Lynch's statements are disconcerting to a scientist. Forecasts are the key to science: we use our conjectured theory to derive mathematical consequences in a newly-designed experiment. If the experimental results disagree, we need to modify the theory and start all over again. The whole process of science, called 'Conjectures and Refutations' by the philosopher Karl Popper, fundamentally relies on our actual experience of reality to be the sole judge of the validity of our theories. But as Niels Bohr said, 'Making predictions is difficult, especially about the future'.

It is the end of the first day of the World Economic Forum meeting, and everyone looks ever so slightly less optimistic than they did last night. The whole day has been about just this — making predictions about how our reality will be in light of technologies that have not yet actually been developed. The words on everyone's lips are the Fourth Industrial Revolution. An event that has not yet happened, but will, soon, at some point, probably. It will be loosely named 'cyber-physical systems', and will undoubtedly disrupt almost every industry. It is all maddeningly hypothetical for a natural scientist such as myself, and it all makes me miss physics a little bit.

It got my fingers on the research button. If the simplest economical transactions take place between two individuals, could it not be that these could at least sometimes be understood quantitatively and subjected to carefully designed and controlled experimentation? Then maybe once we understand transactions between two individuals, we can extrapolate to a large number of individuals. This would be the economics analogues of Bernoulli's derivation (then perfected by Boltzmann) of the behaviour of a large number of atoms in a gas by using the Newtonian theory of collisions of individual atoms. Physics to the rescue!

Unless economical understanding can be underpinned by the same scientific principles as that of natural sciences, the gap between the two would remain forever open. Fortunately, microeconomists studying individual human transactions do approach the topic as physicists would (they like to think of economics as the 'physics of social sciences'). Moreover, there is a discipline that takes the Boltzmann micro-macro approach in physics and applies it to the micro-macro transition in economics. It is of little surprise that this discipline is known as econophysics.

There is one obvious enormous difference that I mentioned before — humans are rather more complicated than atoms. Human transactions, unlike those of atoms, are governed by feelings like hope, fear, and excitement, to name a few. None of this exists in physics (except physicists, of course. We do have emotions, honestly). There is, therefore, frequently a discrepancy between what mathematics predicts an economist would do and what they actually end up doing.

The mathematical foundations of economics were laid by John Von Neumann and Oscar Morgenstern in their classic *Theory of Games and Economic Behaviour*. They invented a whole new branch of mathematics, called game theory, in order to understand economic behaviour. That happens frequently in science. Newton needed to invent calculus (now called higher mathematics) in order to understand the physical laws of motion.

What von Neumann and Morgenstern did was introduce the basic axioms of how rational economists should make their choices in order to maximise profit. The emphasis was on the word 'rational'. Economics, as we said, is the science of how humans make their choices. Rationality implies that the choice always leads to a better outcome for the individual engaged in the transaction. One of the axioms of rationality is therefore the following: if someone prefers apples to pears and pears to bananas, then that person ought to prefer apples to bananas.

But most people do not think like that. If given only two choices, I actually prefer apples to pears and, separately, pears to bananas, but would probably go for a banana if only offered an apple as an alternative. This choice is difficult to understand using the rational basis for economical decision making. As extensive research in behavioural economics shows, humans are nowhere near as rational as von Neumann and Morgenstern have conjectured. Behavioural economics (and psychology) actually observes how real humans behave in economic transactions and try to find mathematical models and various evolutionary explanations for their observations. Can we predict people mathematically?

Two people are asked to split £100, with one condition: one of them decides on the split and the other one accepts or rejects. So, the decider could say take £99 and give £1 to the receiver, or they could decide to go 50/50. The rational theory of von Neumann and Morgenstern suggests that the second person should accept the deal no matter how little they are offered since this is always better than nothing — which was their state at the beginning. It is therefore irrational of the second player to reject the 99/1 split. Yet in practice when this experiment is performed, the first player almost always ends up offering something close to 50/50 and the second almost always refuses if the split is very unequal (70/30). Yes, this is different to how we should rationally approach the problem, but at least there is a proven mathematical solution.

It seems as though humans are hardwired for fairness and reciprocity. This makes perfect sense evolutionary speaking since our survival very much depends on our cooperation with others. So much so that we have a strong feeling of revulsion at transactions we perceive as unfair and a hard-to-resist desire to punish those we perceive not to be playing fairly. But it does put a spanner in the works when it comes to rationality, which is why behavioural economics has to take this into account. Does our morality compromise our rational behaviour, and is this therefore the reason behind the big gap between micro- and macroeconomics?

Psychologist Daniel Kahneman received his Nobel Prize in economics for pioneering this kind of experimentation that ultimately led to behavioural economics. He developed a model of human decision-making that is meant to explain departures from rational behaviour. This model is very akin to Plato's chariot allegory.

The allegory is meant to be a representation of the human soul. Plato says, 'First, the charioteer of the human soul drives a pair, and secondly, one of the horses is noble and of noble breed, but the other quite the opposite in breed and character. Therefore, in our case the driving is necessarily difficult and troublesome.'

The charioteer represents intellect, reason, or the part of the soul that must guide the soul to truth. One horse represents rational or moral impulse or the positive part of passionate nature (e.g., righteous indignation), while the other represents the soul's irrational passions, appetites, or concupiscent nature. The charioteer directs the entire chariot (soul), trying to stop the horses from going different ways, to proceed towards enlightenment.

In Daniel Kahneman's model, there are two systems, imaginatively called system 1 and system 2. System 1 is the impulsive part of our brain that makes quick decisions when encountering complex problems. System 2 is the slow, deliberate, rational one that von Neumann and Morgenstern imagined every economist would be. The difficulty is that we need both, and making sure that we respond appropriately requires us to negotiate between the two (just like Plato's charioteer).

System 1 is instrumental to our survival. When someone shouts fire, we instinctively run outside into a safe zone, often following others doing the same. It would be foolish to waste time by engaging system 2 and trying to estimate all factors, such as the chances of your work catching fire, the credibility of the person shouting and so on. Our ancestors who did engage system 2 when their fellow-tribesmen warned them of an approaching lion surely went extinct. We are the descendants of those who made quick decisions when faced with danger.

System 2, on the other hand, engages when we have to solve tasks such as adding two large numbers, say 345 + 667. It is also at play when we are making big decisions, such as choosing which school to send our children to (or at least it should be). To understand the ability, integrity and strength of the headteacher takes time and rationalising. If we made a snap decision based on chatting to her for a few minutes, noticing that she is charming, witty and good-looking, we would not be making the most considered choice for the future of our children. And yet, very often in human behaviour, system 1 is employed when we would have been better engaging with system 2. I wonder if it is our tendency towards system 1 that makes humans so hard to predict. I wonder if the technological age and our era of immediacy (particularly in the west) makes rational behaviour even scarcer.

Some situations are more complicated than we would like them to be, so we take the easy option of using the rules of thumb underpinning system 1, rather than wasting time, energy and other resources required by the proper use of system 2. This leads to mistakes. A pencil and chewing gum cost £1.10. The pencil is £1 more expensive than the gum. How much does the gum cost?

Most people answer 10 pence. This is the lazy answer that looks obvious to system 1. But it is wrong, since if the gum is 10 pence then the pencil has to be £1.10, and the sum therefore exceeds the total of £1.10. This reasoning is typical of the rational system 2. It takes longer, it is more critical but also more reliable. It is Plato's horse that 'is noble and of noble breed'.

This is a small mistake, because only 10 pence is at stake. However, when we choose our mortgages, or say our leaders, the danger is real and we should guard against using system 1 excessively. It ought ideally to be moderated by system 2. In this way, macroeconomic systems may also suffer from human irrationality. And so, the big gap in economics, the difficulty in using individual transactions to explain macro patterns, comes down to human fallibility.

Inspired by economist John Keynes (and one of the most influential economists of the 20th century), I played the following game with my daughter and my son. I told them to think of a number between one and a hundred. The winner is the person whose number is closer to half of other person's number. To motivate them, I offered £10 to the winner. If they think of the same number, they each get a fiver.

My daughter thought out loud straight away. She said, 'Daddy, can the number be one?' I should have just said 'yes', but I asked why. She explained that whatever the other person imagines, one will always be closer to the half of the other number (think if the other person imagines 100, half of it is 50 and 1 is closer to 50 than is 100, then work your way down). The draw occurs only if the other person imagines the number 1 too, but in this case, they both share £5. Win-win. I was delighted that she figured this out very quickly, but this automatically made my son claim that he also imagined 1 (smart) and so my daughter was forced to split the tenner.

However, imagine now that there are more players, say three. Imagine one of them follows the logic of my daughter and imagines the number 1. One of the other players could just not have figured this logic out and randomly chooses number 50. But now the third player is better off with any number less than 50 and more than 1. In that case, the third player wins a tenner instead of splitting a fiver if they imagined number 1 too.

The game is now more complicated than the two-player game as one has to anticipate the logic of other players. Will they figure out that number 1 is the logical number to imagine? Or will some imagine larger numbers? In that case, I am better off thinking about a number bigger than 1 even if I understand that 1 is the most logical conclusion. But other players might be using the same logic and trying to second guess me. So, I have to think about what they are thinking about what I am thinking about what they are thinking about and so on ad infinitum…Phew. There is a reason I get migraines playing poker.

This illustrates the intrinsic difficulties with many-player game theory as a rational enterprise and the way things unfold in practice. Incidentally, when the number game was played for real in experiments, the winner is usually the person close to the number 25. This means that most people did only one round of the back and forth anticipating thinking. Namely, they thought the other person might go for 50 (half of 100) which means they would be better to go for 25. It did not occur to the majority that other might be thinking the same in which case you should continue halving your guess (until everyone arrives at the number 1!).

Keynes used a similar logic to illustrate how successful stock brokers have to think. He used the analogy of a game where you have to choose a girl among a hundred girls, and the winner is someone who chooses the girl chosen by the majority (this may sound a very shallow experiment, if it weren't for the fact that the man was gay). Would you really choose the girl you think prettiest? After all, looks are of course a matter of taste. What you need to do is choose the girl most other people think is the prettiest. But all other contestants are thinking the same. You have to anticipate what others think you think others think you think…is the prettiest girl.

When investing, therefore, you are trying to pick a stock others will also pick. This is not necessarily the best stock, but it is the one that others would pick as best (given that they too are trying to second guess everyone else). This is a well-known herding behaviour of social animals like us, but can it be made into a mathematically rigorous theory?

In game theory, the stag hunt describes a conflict between safety and social cooperation. Two individuals go out on a hunt. Each can individually choose to hunt a stag or hunt a hare. Each player must choose an action without knowing the choice of the other. If an individual hunts a stag, they must have the cooperation of their partner in order to succeed. An individual can get a hare by themself, but a hare is worth less than a stag. There are many variations of this, known as the 'assurance game', but the original stag hunt dilemma is as follows: a group of hunters have tracked a large stag, and found it to follow a certain path. If all the hunters work together, they can kill the stag and all eat. If they are discovered, or do not cooperate, the stag will flee, and all will go hungry.

The hunters hide and wait along a path. An hour goes by, with no sign of the stag. Two, three, four hours pass, with no trace. A day passes. The stag may not pass every day, but the hunters are reasonably certain that it will come. However, a hare is seen by all hunters moving along the path. If a hunter leaps out and kills the hare, he will eat. However, it results in the trap laid for the stag to be wasted, and the others will starve. There is no certainty that the stag will arrive; the hare is present. The dilemma is that if one hunter waits, he risks one of his fellows killing the hare for himself, sacrificing everyone else. This makes the risk twofold; risk the stag never coming, or risk another man taking the kill.

One reason for this dilemma, similar to the prisoners' dilemma we met with in the prologue — indeed maybe the key reason — is that not all the parties involved in trading have the same amount of information. This leads to the problem of asymmetric information in economics, the solution to which led to three important Nobel Prizes (I know I keep banging on about Nobel Prizes, but come on, these ultimate rewards of life-changing breakthrough are pretty staggering).

The field was introduced when one of the three, George Akerlof, wrote a paper called 'The Market for Lemons'. This paper was thought to be so revolutionary that it was rejected by a number of journals before finally appearing in print. Like with all simple but profound results, the rejecting editors suggested that the result is either obvious or that it will lead to doing economics completely differently (the latter being true, but a funny logic with which to reject a truly new insight). The logic he used to explain his theory was this: Imagine a used car salesman — he knows much more about the cars he is selling than even the most informed buyer. Imagine that half of the cars he is selling are lemons (i.e. in a condition below the selling price). Say that they in practice cost £10k. The other half of the cars are decent and sell for £20k. He is offering the cars at the average of £15k, but you — the buyer — do not know if the car he is telling you about is a lemon or not. He, on the other hand, does.

You can already see the problem. The temptation for the salesman is to put only the lemons on the market. That way he makes the maximum profit of £5k per car sold. But all the buyers lose out and the word spreads and ultimately no one wants to buy used cars since there are no good ones on sale. The market fails, because of asymmetric information. The buyer knows too little and the salesman knows it all.

The market fails, but the scientific methods still succeed in analysing this scenario and offering solutions. Even a somewhat complex situation like this can actually be analysed the way a physicist would do it. Two solutions can be offered and each in themselves a Nobel Prize. They are mirror images of one another and work hand-in-hand. The first is 'signalling', where information is balanced by both parties giving ...

You are on a date. You want to prove (signal) to the person you are dating that you are good solid partner material: trustworthy, loyal, generous, practical (and anything else you deem important). You know these things, but how do you convince your date that you are not selling lemons? I know a lot of really great people who are terrible at dating. The reason? That they are awful at signalling what is great about them! They give a terrible impression: signal failure to the extreme.

I am not saying I know it all (far from it — considering I spent my younger years with my head in physics books instead of dating), but for what it is worth, here is Vlatko's dating advice using signalling where information is asymmetric.

Ask questions (signals that you are interested and not utterly self-absorbed): Do not just talk about yourself and your great achievements. Be curious about the person sitting opposite you — they have their own story, and everyone has something to teach. Everyone likes it when the person talking to them genuinely wants to know them. Another key: listen to the answers. Make an effort to remember little things they may have told you previously. Bring up past jokes. Pay attention.

Be confident (signals that you are capable and trustworthy): Ok, maybe your life is falling apart, you have lost your cat, you cannot get your latest paper right, you spilt red wine all over your crème carpet last night. Your date is not there to listen to your problems. It is not lying, but shelve them for now. Have fun. Laugh. Talk about what is going right, not wrong. I am not saying you should not be honest, but first impressions are important. Select the positive parts of yourself and show them clearly.

Show your skills (signal your strengths): Talk about what you are interested in, what gets you going. Mention recent projects you have done — building your own garage or whatever it is that shows your capabilities. Show your enthusiasm and passion for whatever it is you are into.

Pay (signals your generosity): Take care of the bill without question. It shows explicitly how valuable your date's company is to you.

What if you are not any of these things? Well, you should probably start working on yourself first, but you can try faking it, if you are a decent actor. In this way, you are the lemon, and the information is asymmetric because you are holding the knowledge that you are a lemon from your date. But, mathematically, the truth will out in the end.

Is there anything your date can do to make sure you are not a lemon? Yessiree. That is where 'screening' comes in (the third Nobel Prize in economics). Let me take another example. Now you want a job, not a date (perhaps the latter is contingent on the former, although in my experience — not always). How can your future employer really trust that you are hardworking, resourceful and committed? I mean, you could fake all that at the interview, just as you could on your date.

How can an employer screen for lemons? If they are skilled at interviewing, they can tell integrity from dishonesty by asking the right questions, and understanding genuine behaviour. A successful interviewer screens from the start to the end. Even before the interview, they would use a list of prerequisites to screen. One prerequisite could be a difficult degree. Who could acquire this unless they are genuinely hardworking, resourceful and committed? By doing a hard degree, you are signalling all these attributes, while at the same time, your employer, in asking for it, is screening.

The end of this story of asymmetric information means bad news for the fakes. The best strategy in life is, pure and simple, and no matter how corny it may sound: be yourself. And the reason is simple and can be mathematically justified. If you fake it for long enough you will be found out. And life is a long-distance run. You might win a few battles by faking it, but you will lose the war.

So, asymmetric information is the biggest difficulty in transactions as groups of people get larger. By sticking to certain rules, asymmetric information can be mathematically balanced. Sun Tzu was a very powerful Chinese military strategist and philosopher. Military strategy seems one of the most complex human social behaviours. When two large armies meet, so much seems to be dependent on the random human (and natural) element. What if Hitler had decided to invade Russia earlier? Would he have succeeded simply because the weather conditions would have been more favourable?

Sun Tzu argued to the contrary. He said that there are some strict rules to adhere to if you seek military success. Some of his famous pieces of advice are: 'The supreme art of war is to subdue the enemy without fighting', and 'Know thy self, know thy enemy. A thousand battles, a thousand victories'. It is clear that there are things we can do to be better prepared and informed when we are fighting a battle. And this can only help increase our chances of success.

But can Sun Tzu's rules be formalised into a set of equations that would describe all military conduct, and therefore large group behaviour? Take Niccolò Machiavelli's *The Prince* as another example. This is a manual for how to seize power and then maintain it. Machiavelli gives us some basic principles of what the aspiring ruler — the prince — must have in order to become successful. Each time a rule is broken, Machiavelli gives us a historical example of how the prince in question actually failed.

Could this be a physical approach to power? These are Machiavelli's principles:

- to put down powerful people
- not to allow a foreign power to gain reputation
- to indulge the lesser powers of the area without increasing their power
- to install one's princedom in the new acquisition, or, even better, to install colonies of one's people there

This is not vastly different to how Sun Tzu would advise. Although these rules are not exactly mathematical, it is claimed (and sadly proved over and over again through history) that deviating from them leads to failure. This is an example of reductionism, which may not be mathematical, but is still a viable option of closing this big economics gap.

Let me introduce a more modern reductionist, who belongs to the Austrian school of economics (a type of economics that believes all social phenomena results from the actions of individuals), Friedrich Hayek. He would probably not be considered a reductionist, but I think he is one *par excellence*. His most popular book, *The Road to Serfdom*, is definitely a logically-argued sequence of consequences that result from a society deviating from the philosophy of individualism. He takes individual freedom to be paramount and shows — very logically — how a society that starts to gradually erode individual freedom must ultimately end up being a dictatorship of a fascist or a communist type. I think Hayek actually saw little fundamental difference between left wing and right wing dictatorships. I always remember the memorable words of my once-neighbour in Vienna, the Nazi hunter Simon Wiesenthal, who summed this up as: 'you go left, left, left and you end up on the right'.

Hayek appeals to many people on an emotional level. I like the idea that individual freedoms should come before other freedoms, but what I like even more is that he tries to appeal to our reason that this should be so. His book comes as close as I have seen to arguing how a natural scientist would argue what happens if certain core principles are abandoned. True, his book is no *Principia*. There are no equations written to characterise a good human society (even though this is his sole purpose). In fact, as I understand, he and the school of thought he belonged to would be against the idea that human social behaviour could ever be captured using any mathematical formalism. In this, I believe, they were very much mistaken. In fact, I would go as far as suggesting that this could be how the gap between micro- and macroeconomics could be bridged.

I can obviously see why macroeconomic systems seem far too complex to be explained by free transactions between individuals. But surely this is where the beauty of economics truly lies. Simple systems, almost by default, are too easy to understand to warrant immense interest. If the expansion and contraction of the housing market alternated between boom and bust with every incoming week, then who would need an economist (or a journalist)? Any simpleton could make the right investments by tracking whether the current week is boom or bust — a feat that requires remembering no more than single bit of information — a single 0 or 1.

Real world systems are all vastly more complex. The actual economy consists of a mindboggling network of interconnected components, from national policy, consumer outlook, to civil unrest. Perturbation in one sector can have immense consequences — as witnessed during our most recent financial crisis where banks were bailed out since they were simply 'too big to fail'. This culminates in a rich tapestry of behaviour — requiring us to keep track of far more than the single bit of information involved for the simple oscillation between boom and bust.

The non-triviality of complexity is also a blessing. A world that operates in predictable 'boom-bust' cycles, depending on single bits, would be a rather boring world. Complex systems form a basis of our society — whose different institutions — schools, courts, factories and so on — have to work together to achieve a state beyond the simple hunter-gatherer society. Meanwhile, life itself demands complexity — cells in our body work in unison to keep ourselves functional. Without complexity, life could not exist. The fact that our universe turns out to be complex is probably very fortunate for us.

It is the very reason I am here in Dubai. It is no surprise that the prediction and understanding of complex systems is of great interest for the World Economic Forum. It revolves around the question of how to accurately predict and mitigate what is coming next, whether financial, humanitarian or ecological. Our capacity to answer these questions may be crucial for our survival. The Fourth Industrial Revolution that is our topic du jour, sees great potential for society. The previous Industrial Revolutions (steam in 1784, electricity in 1870, information technology in 1969), did good things for society as a by-product: the UN was formed, many colonised people in Asia, Africa and the Middle East won independence, the Civil Rights Movements began, to name but a few. Understanding complex systems is vital for our survival as a human race.

The fundamental principle that captures the mathematical understanding of complex systems is that of cause and effect. We record information from the past and make use of it to gain greater knowledge — and thus benefit — in the future. We see this in the movie *21*, where a group of MIT graduates take on the casinos of Las Vegas via the game of black-jack. Here, going into the game, oblivious to what cards were previously dealt, will always result in a house edge. Yet, by tracking this information — a player can better predict his future odds of winning — and twist the game in his favour.

Modelling generic processes is a similar game; by recording past information, one is better equipped for the future. In the toy system of the oscillating stock market, this is simple. If we knew a bust one week always precedes a boom in the next, then a single bit of information (whether the current week was a boom) gives us all the necessary information for optimal investment. The game of blackjack is more complex, but standard card counting strategies are feasible for sharp minds — one counts how many 10s and aces have been dealt compared to the cards numbered 2-6. The systems studied at the World Economic Forum are, of course, far more complex than games in the casino — the amount of relevant past information is immense. Indeed, the minimal amount of past information one must record to make optimal predictions is a well-known quantity of complexity in the scientific community — known to mathematicians as the statistical complexity.

The value of quantum theory for the understanding of such complex systems may at first sight appear paradoxical. Whether it is a game of cards or the institutions that underlie the economy — all concepts of relevance are merely classical. How can quantum modelling possibly help?

It turns out that for many processes, even the optimal classical models are wasteful — they require information from the past that has no bearing on the future. Take, for example, a simple variation of the boom bust economy. Instead of a guaranteed boom after every bust, let us supposed it was simply likely — say with probability $p = 0.8$. Let X now be the variable that represents whether the previous week was boom or bust. It is clear X is a cause of future events, and thus must be recorded to make beneficial future investments. Nevertheless, even if we are to observe the entire future, we would not be certain whether the previous week was boom or bust. Some parts of information we store is never reflected in future statistics, and is thus wasted. The potential consequences go beyond gaming casinos. In 1961, physicist Rolf Landauer showed each bit of wasted information also incurs extra energy — information wasted is energy wasted. Discussing environmental heating due to computation has been a big topic of the day, I am sure you would not be surprised to hear.

Quantum logic offers improvement. X need not take on a definite value. We may store the conditions 'X = bust' and 'X = boom' in a super-position of '0' and '1'. Thus, we can be in a superposition of recording and forgetting whether the week prior was boom or bust — making a substantial saving on the amount of information required to correctly simulate this simple economy. Perhaps in this way, quantum modelling can offer help in bridging the micro- to macroeconomic gap.

While saving part of a bit may not sound grand, this strategy generalises. The more complex a process, the greater the cost of assuming a definite reality. To a person capable of storing and processing quantum information — the universe could look far less complex. This offers a new paradigm, where our notions of what is complex ultimately depends on the information theory we use.

CHAPTER VI

SOCIOBIOLOGY

I f an economic system is a complex tapestry because human behaviour makes up part of it, what about a social system, made entirely out of the actions of humans? What works on a societal level and what fails? This very topic continues to be fascinating: dystopian fiction is prolific and enthralling often because it predicts what is to come, and it takes a frighteningly small flick of the pen to get from a current problem in society to a full-blown nightmare. It is compelling precisely because we recognise our own behaviour (albeit a diluted form) in the pages.

I am in Brussels right now and staring down the barrel of a long few days on a panel for the European Research Council, listening to funding proposals for every type of science imaginable. I used the flight over here to finish the novel *We*, a *tour de force* by Russian writer Yevgeny Zamyatin. It means I have got nothing to read under the table while listening to the proposals, but when·you get to the last chapter of this novel, you do not stop for anything.

We* is set in the future. D-503, a spacecraft engineer, lives in the One State, an urban nation constructed almost entirely of glass, which assists mass surveillance. The structure of the state is Panopticon-like. The Panopticon is a type of institutional building designed by the English philosopher and social theorist Jeremy Bentham in the late 18th century. The concept of the design is to allow all (*pan*) inmates of an institution to be observed (*opticon*) by a single watchman without the inmates being able to tell whether or not they are being watched.

Furthermore, life is scientifically managed F. W. Taylor-style. People march in step with each other and are uniformed. There is no way of referring to people save by their given numbers. The society is run strictly by logic or reason as the primary justification for the laws or the construct of society. The individual's behaviour is based on logic by way of formulas and equations outlined by the One State.

Obviously, no social structure has ever been as rigid as this, but some, such as present-day North Korea, have come frighteningly close. This is what makes the novel work, what makes the hairs on your arms stand on end as you read. 'Marx was right — he just got the wrong species,' says Edward O. Wilson, the founder of sociobiology. He studied the behaviour of ants extensively, and is saying here that communism describes ants much better than it does humans. They live in communities where their roles are entirely genetically determined. They presumably never question their pecking order or rebel against the social order that is imposed from 'above'. And yet it has never worked successfully with humans, despite many lengthy attempts and despite some great intentions behind them.

Wilson's field, sociobiology, is 'the systematic study of the biological basis of all social behaviour'. It is controversial by its nature as it claims that social behaviour can be understood in terms of biological behaviour. And this is a sensitive topic for us humans. It touches the 'free will' nerve. Ouch. We like to believe that we are able to escape at least some aspects of our human nature, that we have some choice in how we act. Lower species like ants might well be more determined, but surely, we humans are free?

Brussels is an interesting place in which to digest the ideas in *We*, and an apt place to think about people working together on a large scale. Since the end of the Second World War, Brussels has been a major

---

*Please see Wikipedia entry for *We* for full description, https://en.wikipedia.org/wiki/We_(novel)

centre for international politics and has become the home of numerous organisations, politicians, diplomats and civil servants. It is the capital of (do not talk to me about Brexit) the European Union. It is an extremely multilingual city, I am realising sitting outside a café next to Grand Place, the city's biggest and most spectacular square, hearing snippets of conversations drift past. It has welcomed increasing numbers of migrants and expatriates over the last few decades, a real example of tolerance and libertarianism.

Could sociobiology be our missing link? If we can explain individual human behaviour using biology, then perhaps it can also explain not just the large-scale transactions of economics, but also the macro human behaviour of sociology. Let me clarify — it is not that Wilson is the first person to suggest reducing social to natural science. Walter Bagehot, an English Victorian economist, wrote precisely about this in *Physics and Politics*. *Physics and Politics*, unlike the suggestion of its title, is not about reducing politics to physics (though Bagehot must have had Newton's *Principia* in mind as a model of scientific perfection). It is really about the application of Darwin's evolutionary biological idea to the development of society.

An interesting quote from Bagehot (Victorianism aside) follows thus: 'Even some very high races, as the French and Irish, seem in troubled times hardly to be stable at all, but to be carried everywhere as the passions of the moment and the ideas generated at the hour may determine. But thoroughly to deal with such phenomena as these, we must examine the mode in which national characters can be emancipated from the rule of custom, and can be prepared for the use of choice.'

Freed from the expectations and traditions of our customs, how does our choice and free will work biologically? Darwin's evolutionary biology, as we explored earlier, is about adaptation over time, the best-adapted surviving, and different organisms emerging as a result. Is this the same for societal behaviour? What makes us act collectively? The gap in social sciences is all about reconciling the behaviour of individuals with the behaviour of large crowds, like city inhabitants or residents of a country.

I am thinking of another very open-minded city, a little closer to my current home. A city that has graves of the inventors of communism and libertarianism next to one another. I am talking about London, and the famous Highgate cemetery, where both Karl Marx and Herbert Spencer were buried. Marx's grave is possibly one of the most visited in London. It is why people are most likely to come to the Highgate Cemetery in the first place (and as a by-product find themselves in an unrivalled place of beauty and calm).

Marx's views were (broadly speaking) that we should suppress all instances of individuality — social benefits are the only important outcome. Spencer argued for the opposite: there is no such thing as a society — it is all just a bunch of individuals (here I am actually misquoting Margaret Thatcher, but I think you get what I mean). In neither of these views is there any conflict between the individuals and society. That is because they deny one of the two aspects, the individualistic or the social. Most other social views are somewhere in-between. In fact, the libertarians are these days becoming paternalistic (arguing for instance for citizen's wages), behaviour you would expect from communists, and communists are becoming more free-market friendly (just look at the Chinese government). This is a global trend. And maybe for the better. Perhaps we need some kind of right mixture of all of these to find an equilibrium, if such thing can ever be found in a human society — it is obvious that both extremes lead to inefficiencies and moving closer to the middle would come closer to a society that functions better.

I have been spending a fair bit of time in Beijing this year, as part of my sabbatical. Learning more about its history has been an enjoyable accident — it is rather bloodthirsty. The Forbidden City in Beijing was home to two different dynasties, Ming and Qing, and 24 different emperors in total throughout its colourful and turbulent history. It was built in the early fifteenth century, some 900 buildings and 9000 rooms and antechambers.

The architecture of the City is a prime example of Eastern simplicity and beauty. Some of the grandest buildings within the Forbidden City were the residences of the Emperor and the Empress. The rest of the rooms were dedicated to the supporting staff, which included some *one thousand* concubines.

One thousand concubines? Imagine the competition for the emperor's attention? There are stories in the Chinese folklore told about the level of cunning exhibited by some concubines in attempts to orchestrate 'chance' encounters with the emperor that could potentially lead to several nights spent by his side. Having a child fathered by the emperor was like *almost* winning a million on the lottery. The prize money is no guarantee — as a concubine rearing one of the emperor's children (male, since only males could inherit the throne), you must somehow eliminate other older children to make sure your son will inherit the throne. In such a closely-knit 'family', intrigues, jealousy, backstabbing, murders were almost as common as drinking tea.

J. B. S. Haldane, a British-born Indian scientist, joked that he would willingly give his life for two brothers, four nephews, or eight of his cousins. This was then formalised by William Hamilton.

Formally, genes should increase in frequency in a population when $rB > C$ where

> r = the genetic relatedness of the recipient to the actor, often defined as the probability that a gene picked randomly from each at the same locus is identical by descent
>
> B = the additional reproductive benefit gained by the recipient of the altruistic act
>
> C = the reproductive cost to the individual performing the act,

In other words, if genetic benefits exceed the costs of dying.

Even closer to home than all the killings in the Ming and Qing dynasties is the equally bloodthirsty English royal family during the Wars of the Roses. Over a two-hundred-year period (up to the end of the fifteenth century), there were 47 murders in the battle for the crown, all but five of which involved cousins (whose value of r is only an eight). Out of the remaining five, two are brothers, and three nephews. And this does not include the many suspected but never confirmed cases, like Richard III killing his two nephews.

According to Hamilton's chilling logic, you should be prepared to kill up to 4 cousins (or two nephews) in order to save a brother. Also, and harder to swallow, you should be prepared to kill a brother in order to save five or more cousins, or three or more nephews (by say becoming the king by disposing of your king brother). None of the documented murders violate the Hamilton formula above. In fact, regicide in history is a great arena in which to test him.

This kind of kinship is so deeply ingrained in our being that we follow it subconsciously even when there is no direct benefit to us or to our cousins. A recent experiment measured how long people can hold their breath for: it found that most can hold their breath for significantly longer if they are told that they are doing it for their brother than if they are doing it just to prove themselves.

Genetics has a direct effect on many aspects of our behaviour and certainly influences the social structures humans have created. We usually feel a tremendous amount of loyalty towards our family and this extends to very few people outside of this narrow circle. The way we extend the circle beyond genetics to our friends also mirrors the behaviour established on the kinship logic.

So, in the battle against the inevitable slide towards chaos, molecules had to be invented to provide a bit more durability to atoms. These molecules have then given rise to replicating molecules — like DNA — which significantly extends the longevity of structure. DNA in turn has engineered for itself vessels — plants and animals — which would protect it even further against the inevitable degrading. Finally, larger structures have spontaneously assembled — such as packs, bands, families, nations — which extend persistence of certain physical structures well beyond the capability of the individual atoms. Social structures ensure our survival.

Suppose you are an individual in a society where all interactions take the form of the prisoner's dilemma we met with earlier, where you must decide independently whether to cooperate or defect. It may help us understand how large-scale features can arise out of simple two-player interactions. As you interact with other individuals many times, you actually effectively perceive them as a player with an average random strategy sometimes cooperating and other times not. You too have a mixed strategy, sometime cooperating and sometimes defecting.

The following logic explains why the probability that you will cooperate is bound to decrease with time and ultimately end up at zero. Since this is true for every member of society, society as a whole would unfortunately necessarily go into the state where everyone is non-cooperating. The reason for this is that the probability of cooperation changes in proportion to the difference between the cooperation and defection payoffs. Since defection is more rewarding to an individual, the proportion of co-operators goes down in time. Wonderful! Everyone will end up stabbing each other in the back, with only their own interests as their only motivation! What a lovely society this would be.

Is there a way out of this coming true? The answer is yes. Phew.

Good news for us humans. It relies on something we do very well, which is copying other people's behaviour. Looking around Grand Place, most people seem to be cooperating in unison to something unseen. There is not really any behaviour I can see that looks 'out of place' from how the majority are acting. Why? Why copy others, given that defection is more rewarding? The answer is that there is a stable configuration of individuals in which co-operators can form a cluster, so that within that cluster, copying is bound to lead to more cooperation. It is a safe place to be inside, and leads to connection and community. The only weak point of the cluster of co-operators is the boundary with those in society who defect. But even the boundary of co-operators can be stable, simply because they still gain when interacting with people from within the cluster — which are just as numerous as the defectors outside. On average, the co-operators at the boundary actually breakeven — the ones on the inside are win-win when interacting with one another. And the cluster configuration of co-operators therefore remains protected in a society even when faced with many defectors.

Sociobiology is the field premised on the fact that the gap between biology and human society is actually an artificial one. Instead, the structures arising in all human societies are actually just a natural extension of genetics, and therefore genetic. It is not hard to see why. Genetics govern not only our physical makeup, but also our psychological profile. How we behave will therefore be selected on the genetic basis — the same as the colour of our hair or the shape of our nose. Early structures of human societies, bands of closely-related individuals between 50 and 200, are therefore directly determined by human genetics (or so sociobiology would suggest). Some of these early societies survive, most do not, and therefore biological selection works the same way to ensure the 'survival of the fittest societies'.

Wilson boldly asserted that 'The intellect was not constructed to understand atoms or even to understand itself, but to promote the survival of human genes.' He also noted with fervour the 'growing awareness that [religious] beliefs are really enabling mechanisms for survival.' Morality, too, is just 'another technique by which human genetic material has been and will be kept intact.'

Many sociobiologists study the behaviour of animals as well as humans. Animals are often social beings too, sometimes on par with or more so than humans. Newly-dominant male lions will often kill cubs in the pride that were not sired by them. This behaviour is adaptive in evolutionary terms because killing the cubs eliminates competition for their own offspring, causes the nursing females to come into heat faster, thus allowing more of his genes to enter into the population. Sociobiologists would view this instinctual cub-killing behaviour as having been inherited through the genes of successfully reproducing male lions, whereas non-killing behaviour may have died out as those lions were less successful in reproducing. It is not really so different from the Qing and Ming dynasties, or the English royal families during the Wars of the Roses.

One book that has been hugely influential in attacking the cooperation issue is *The Evolution of Cooperation* by Robert Axelrod. Despite the absence of equations, he shows how we can use exactly the same mathematics that are used in physics and in biology to understand cooperation. When linking Axelrod's book to Richard Dawkins' *The Selfish Gene*, which claims that underlying human behaviour is this selfishness of genes in some sense, then any cooperative behaviour becomes a mystery. Why do we ever actually cooperate with each other? Why do we have this built-in? And why do societies or tribes cooperate with other tribes and other societies? Axelrod really tackles this in his book, which generated a whole field of taking the game theory of mathematics and trying to apply it to social behaviour in order to understand conflict and cooperation.

An interesting idea of his. even if you have an underlying selfish tendency, then cooperation can evolve simply because you are forced to interact with someone else over and over again. If you interact with someone just once, then there is no incentive to cooperate. But if you know that you will be interacting with a person over and over again, where you can check and verify what the other person is doing, and, crucially, if you do not know how long this interaction will last, then somehow mathematics would suggest that it is better for you to switch to cooperation rather than to continue to be selfish. He goes through lots of computer simulations and also experiments with people and some animal species to show that cooperation can evolve. So, in a way it is a very optimistic book in that sense: although our first instinct is often to protect our own interests, it seems that evolution would really favour cooperation. After all, is connection not what all humans are really looking for, deep down and before everything else?

Last night I went to see the opera Cappricio at the Palais de la Monnaie at the invitation of a colleague of mine. It was a wonderful performance, but I found myself with one eye on the audience, still thinking of social behaviour, noticing how applause is another remarkable example of social self-organisation. At the end of a good performance, the audience, after an initial uncoordinated phase, produced synchronised clapping *every time*, where everybody claps at the same time and with the same frequency. Synchronisation occurs in many biological and sociological processes, from the flashing of Southeast Asian fireflies to the chirping of crickets, from oscillating chemical reactions to menstrual cycles of women living together for long periods of time. Rhythmic applauses have been explored in detail, both empirically and theoretically. In the first pioneering investigations, applauses were recorded after several good theatre and opera performances, with microphones placed both at some distance from the audience and close to randomly-selected spectators. The intensity of noise, heard at some distance from the audience or by one of the spectators shows that the signal becomes periodic during the phase of the rhythmic applause, and that the average intensity of the sound decreases. This is not happening consciously — we are coordinating naturally, independent action becoming herding behaviour on its own.

If the fire alarm had gone off during the performance, we would have seen an example of independent action versus herding behaviour played out again. If everyone acts independently, the panicking crowd with its excessive body compression and tangential motion would make it very difficult for people to leave a room through a single narrow exit. It would generate intermittent clogging of the exit, so that people are unable to flow continuously out of the room, and instead, groups of individuals would try to escape in irregular succession. Because of the friction of people in contact, the time to empty the room is minimal in correspondence to some optimal value of the individual speed: for higher speeds, the total escape time increases (the 'faster is slower' effect). If there had been a real fire, meaning people needed to escape from the smoky theatre hall, unable to see its exits unless one happens to stand close to them? In this case, they would not have a preferential direction of motion, as they have to find the exits first. Would it be more effective for the individuals to act on their own or to rely on the action of those close to them? The process is modelled by introducing a panic parameter that expresses the relative importance of independent action and herding behaviour. It turns out that the optimal chances of survival are attained when each individual adopts a mixed strategy, based both on personal initiative and on herding. Luckily, the night ended without drama. Until we went to a whisky bar, of course.

Something that definitely did not happen during this rather highbrow performance was the Mexican wave (also called La Ola) — another good example of coherent collective motion. The model of this peculiar phenomena is inspired by the literature on excitable media, where each unit of a system can switch from an inactive to an active state if the density of active units in their neighbourhood exceeds a critical threshold. The influence of a neighbour on an excitable subject decreases with its distance from the subject and is higher if the neighbour sits on the side where the wave comes from. The total influence of the neighbours is compared with the activation threshold of the spectator, which is uniformly distributed in some range of values. It turns out that a group of spectators must exceed a critical mass in order to initiate the process. The models are able to reproduce size, form, velocity and stability of real waves.

CONCLUSION
Can We Bridge the Social-Natural Science Gap?

P erhaps you remember me mentioning early in this text the rather provocative statement of Ernest Rutherford, who said that 'Science is either physics or stamp collecting'. What he meant by this is that other, more complex sciences than physics, appear to be more like a collection of known facts, rather than a systematic understanding of phenomena that can be logically related to one another. But these 'more complex sciences' have advanced a great deal since the time of Rutherford. The momentous discovery of DNA gave further support to the theory of evolution that there is a common thread to all life. All complex life can therefore be reduced to simpler forms of life. Social science, or, human behaviour, can quite easily be paired with the natural science of biology. So where does physics come into the picture? Even the most optimistic reductionist may be sceptical as to whether the social sciences can be 'reduced' to physics in the same way they can be explained by biology. Can physics actually help us to understand social behaviour, in the same way that biology can?

Often, reductionism takes the form of understanding phenomena treated in one theory using the language of another one. A strong motivation for this is simply to reduce all our understanding to one set of principles and provide a common unifying language to all sciences. However, the real bonus with the reduction about which I will speculate now, is the hope that we can explain something that seems haphazard in one theory with a deeper logic that might help explain it in a simpler way.

At the more speculative end of this journey, we talked about life not only conforming to the physical laws of thermodynamics, but maybe even that the very appearance of living systems — that is currently a central mystery in biology — can, in fact, be just a consequence of the need (in terms of the laws of physics regarding work and heat dissipation) for more thermodynamical efficiency.

And, yes, perhaps there are questions we may never be able to answer with the present approach, so long as the assumptions of Gödel and Turing hold. But not being able to explain the nitty-gritty of chemical dynamics may not prevent us from understanding life with the microscopic laws of quantum physics. Could the same be true of the gap between the natural and social sciences?

I once appeared in a BBC radio interview hosted by Tim Marlow, and to my surprise, the other interviewee was the well-known historian Niall Fergusson. I have always been interested in his ideas, which are on the controversial side, since I read a few of his books in my younger years. Perhaps most controversial are his liberal economic arguments that the British Empire was largely a positive enterprise — so much so that he argues America ought now to take over properly and rule the world using the same British blueprint. True, I am simplifying his complex arguments a bit, but not by much. I was surprised to be sitting opposite him as our subjects do not immediately appear to correlate. I certainly did not expect to get much sympathy from him, but I did not mind much either. After all, I had the usual advantage: everyone thinks (mistakenly) that quantum physicists are very brainy.

At some point, I deliberately said that physics in my view is the most fundamental human search for knowledge. I looked at him in a provocative way, expecting him to disagree and claim that history is far more important. But to my surprise, he nodded. 'Actually, I agree with you. My mum is also a physicist.'

Conclusion

Along our journey, we talked of whether the gap between quantum physics and gravity could vanish if gravity is understood as a thermodynamical quantum effect. If all the jitters of all quantum fields in the universe are added up, there is enough energy there to account for gravity. So maybe gravity is not a fundamental force, and therefore, there is no need to quantise it. A clear prediction of this is that gravitons will never be detected. If they ever are, this idea fails.

We also saw that many thinkers have speculated on the possibility that life itself is also a consequence of thermodynamics. Some would put this more strongly, namely, given the right conditions, life is actually a thermodynamical necessity. This means that if we subject a lump of matter to a strong energy flux, this lump will sooner or later become alive and start reproducing. We do not quite understand fully all the needed conditions and how exactly to define life, but the idea sounds quite appealing. At the risk of sounding like Victor Frankenstein before the conception of his monster, the gap between living systems and inanimate matter would simply vanish.

The idea that life is necessary is appealing for a whole host of reasons, including the old 'purpose question' that plagues many of us. Appealing, yes, but can it be tested? According to the standard theory of evolution, new mutations get selected by the survival of the individual carrying them. Those that do not tend to die out. However, if thermodynamics was responsible for mutations too, then we could imagine that engineering the right external conditions, like shining the right kind of laser light, could also induce desired mutations. An experiment like this would still be challenging, as all other sources of mutations would have to be eliminated. So we would have to show that there are no other evolutionary pressures to mutate and that it happens only because of thermodynamics!

Could it be that various social structures are also a form of driven, dissipative self-organised systems? Namely, given the right conditions, such as climate, geography, ground, and so on, could it be that our social structures also emerge spontaneously? After all, just like complex organisms spontaneously organise by a conglomeration of simpler ones who symbiotically benefit from one another by cooperating together, humans likewise cohabit together and merge into larger social structures for their mutual benefit. Human society, according to this simple logic, is just an even larger form of a living organism.

Most parts of the universe do not have the right conditions to force evolutions of life. The entropy production afforded by ordinary non-living matter suffices. Other parts, like the earth, have a continuous strong energy intake that forces inanimate matter to organise itself more efficiently in order to maximise the entropy production. This leads to life and structures that life gives rise to. Each next level corresponding to an even higher degree of complexity and organisation needs to address the external thermodynamical conditions.

Interpreting anything presents us with three options. One; the thing is like that intrinsically, two; it is because of the outside context and influences, or three; because that is the way we perceive it. There is a parable in eastern philosophy of two students arguing about the waving flag they are both observing. The first says the flag is waving beautifully. The second says, no, the wind is blowing and the flag is just following. A monk walks by and overhears the dialogue. 'No,' he says, 'both of you are wrong. Neither the flag nor the wind are moving. It is your mind that is moving.'

Often, our perceptions mislead us into believing something is true. 'Just because nobody complains does not mean all parachutes are perfect,' said Benny Hill.

This morning, on the panel for the ERC, we listened to an impressively passionate proposal for research into MRI. MRI, Magnetic Resonance Imaging, is a physics technique where strong magnetic fields are applied in order to test the magnetic response of the material under investigation. You can think of each atom in any substance as a little magnet that orients itself — like the compass needle — in the direction of the external field. If you make the external field rotate, the atomic magnets will rotate too. This kind of magnetic manipulation of atoms was pioneered by physicist, Isidor Rabi. He first called it Nuclear Magnetic Resonance before it became the imaging technique we are more familiar with, very useful in medicine ('Nuclear' was dropped from the name for fear of scaring the public!).

While MRI was good to measure the static response, a derivative of this technology called 'functional' or fNMRI, was invented to measure the dynamics in the substance. When applied to monitoring the brain activity it actually measures the blood flow around the brain which tells us which areas of the brain are active — the blood flows to where energy (oxygen) is needed. Without pretending to be a neuroscientist, the two areas I know about are the anterior insula where emotions are processed (yes, even in us physicists), and the dorsolateral prefrontal cortex, responsible for cognition. Through fNMRI, we can see which parts of the brain are working more or less when faced with decisions. Brain imaging allows scientists to build a catalogue of brain areas and their functions, which can then be cross-referenced with behaviours that employ the same processes, says Jonathan Cohen of Princeton. Eventually, he argues, this combination of behavioural analysis and biological neuroscience could inform questions in fields from philosophy to economics. The current study, he said, 'is a really nice example of how cognitive neuroscience — and neuroimaging in particular — provides an interface between the sciences and the humanities.'

It is easy to see how neuroscience bridges between natural and social sciences in this way. Experiments can monitor your brain activity while you are making economic and social decisions. The question is: are you deciding rationally or emotionally? In other words, are you responding instinctively using system 1, or are your actions guided by the slower and more deliberate thought-processes of system 2? fNMRI can be used to monitor the brain activity during the 100 dollars game we saw earlier. This is the game of two players, where one of them holds 100 dollars and decides how to split it with the other player. The only power the other player has is to accept the proposed split (in which case, this is the outcome) or reject it (in which case neither player gets anything).

The finding, using fNMRI, is that if the split is very unequal (70:30 or more) the second player will quickly reject it. The fNMRI suggests the system 1 is at play. We are hardwired to find unequal deals unfair. This is despite the fact that, rationally speaking, we should accept any split given that we are starting with nothing and we always set to gain something unless the first player decides to keep all 100 dollars (even in this case we do not lose anything). But, presumably because humans have evolved in tightly-knit communities, we are hardwired for reciprocity. We expect others to give us an equal, or nearly equal, share. This is something that is encoded in system 1, possibly during the times when we were hardly distinguishable (if at all) from other apes. Reciprocity and cooperation are a hallmark of any social species, and we are more sociable than most.

It is not a surprise that we expect others to be kind to us. All religions have it: do unto others as you would have them do unto you.

There is a possible saving grace of our irrational behaviour: that our irrational response actually serves a bigger rational purpose. We have been exploring how physics connects to chemistry connects to biology, then neuroscience to economics and psychology. In this way, quantum physics is indirectly connected to human psychology. But is there a more direct route? Meet the beautifully-named theory of quantum psychology.

There are many instances where tiny quantum effects are amplified to macroscopic scales. For instance, the radiation from the sun or more or less any other object reveals to us that light is made up of tiny particles, called photons. And this comes by measuring the macroscopic effects of radiation without ever having to detect single photons (which was impossible in Einstein's time, but something we nowadays easily perform in the lab). The ways that large solids absorb heat, conduct electricity and respond to the external magnetic field actually all betray their underlying quantum nature.

So the big question is: if inanimate large objects reveal their quantum nature at macroscopic level, why not living systems like ourselves? Could some aspects of human psychology actually be *dependent* on quantum physics?

The first time this was clearly suggested was by Niels Bohr — he used it more as an analogy but some of his writings suggest that he took it more seriously than just that. Bohr's complementarity suggests that there are some features of the world that cannot be revealed in one and the same experiment. In the double-slit experiment, where photons are fired through two slits to reveal either particle or wave behaviour, we can show that the electron is a particle, or, in a separate experiment, that it is a wave. But there is no experiment that would demonstrate both wave and particle natures simultaneously. Being a wave or a particle are complementary features. When we get a click in the electron detector, we confirm the electron's particle nature. We can then get another click later on, confirming again that the same electron is a particle. However, in-between these two clicks, quantum physics says that the electron must have been a wave. In other words, the electron must have taken many different routes from initially being a particle to finally being a particle.

146

There is no way of explaining occasional particle behaviour without admitting that in-between electrons have to be waves.

Bohr saw this kind of complementary at work in many different situations, not just in physics. He would say that the conscious and subconscious parts of our brain are in fact similarly complementary. When through introspection we arrive at a definitive conscious state of mind (the psychological equivalent of being a particle), there is no way to use introspection to arrive at this state of mind through other definitive states. Prior to a definitive state, our brain must have been more like a wave, fuzzy, indeterminate and uncertain. The subconscious, in this view, is a direct and necessary complementary counterpart to being conscious. It has even been suggested that the fuzziness of dreams stems from the same, necessary, wave-like quality of the subconscious.

Show me the proof! I hear you cry. There are two levels at which this question could be tackled experimentally. They are both speculative, but — I believe, possible and even probable. The first is at the micro level in the brain where experiments are being done to probe and see if quantum wavelike behaviour both exists and plays a role in some fundamental processes in the brain. The most progressive experiments to date in this area are proposed by Matthew Fisher, whom we will get to know a little later. The other level of experimentation is, shock horror, macroscopically, where studying human perception could reveal quantum aspects in the brain. This time, we will begin macro and then look more closely into Fisher and his theory. Necker cube, I welcome you to the stage!

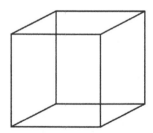

This is a two dimensional line drawing of the cube first brought to light by Swiss psychologist Louis Necker. Even though there are many logical interpretations of the picture (many actually impossible in 3 dimensions) the brain does not see any ambiguity in this two dimensional drawing. In fact, we see only two of the possible images (with one of the surfaces at the front or at the back). Most people see one of these first (this is apparently to do with the way we understand perspective) and then, after a few seconds or so, revert to the other one, thereafter continuing to flip back and forth between the two.

What could possibly be quantum about this? Well, suppose that the two images of the Necker cube are stored in our brain as two distinct physical states. Admittedly, these states could be very complex, in the sense of involving many atoms and interactions between them. The mind then switches between the two physical states, which is like a logical operation of a flip from zero to one and back to zero. The question is: could it be that the switching process is actually quantum mechanical and that during the transition our brain is actually storing a superposition of the two images?

The suggestion that this flipping could actually be quantum was, to the best of my knowledge, first voiced by the physicist Elihu Lubkin. He was inspired by Bohr's statements relating to complementarity and human psychology and wanted to bring it to the experimentally testable level. Other physicists, like David Finkelstein, have also expressed their suspicions that the brain might be using the quantum logic instead of classical.

More recently, Harald Atmanspacher and his colleagues from Germany suggested that what might be at work in the perception of the Necker cube is a subtle combination of quantum dynamics intercepted repeatedly by quantum measurements. This is something that is known as the quantum Zeno effect. The idea is as follows: the encoding of the two images in our brain would quantumly oscillate between the two images of the cube. The natural period of this oscillation is, say, one tenth of a second. But our conscious part of the mind actually measures this quantum state resulting in either one image or the other. This measurement slows down the evolution between the two images (hence Zeno).

We have three distinct timescales here. One is the rate at which consciousness operates. We are not exactly sure how long this is, but the minimum requirement is that we can clearly discriminate two events in time. If two sounds (audio signals) are separated by less than 3 milliseconds, then humans cannot tell them apart. Between 3 and 30 milliseconds, most subjects can hear two sounds but they fail to put them in the right order. The ordering of the events is randomised. Beyond

30 milliseconds, most humans can clearly discriminate two sounds. Psychologists therefore believe that it is reasonable to assume that the duration of the 'now' for humans is about 30 milliseconds as that is how long we hold our perception for.

The second timescale is a third of a second. This is usually taken to be the time we become aware of something. Experiments show that we actually consciously think we do something about a third of a second after we have unconsciously made the decision to do it. So just as the image starts to flip in a quantum way in our mind, it gets measured. Before it is measured, the hypothesis is that it exists simultaneously in two states in our mind. The image is literally in a quantum superposition.

And the third timescale (we're working with lots of threes here), is three seconds. Because of the measurement by our consciousness, the time of switching between the two images gets longer. A simple quantum argument shows that this ends up being on average three seconds. And the Necker cube? Evidence shows that most people report a flipping time that agrees with this. It is right there — try it for yourself.

There are many things humans do with the rough period of three seconds. For instance, the length of lines in song verses is also approximately three seconds, as well as melody phrases (Deep Purple's riff 'Smoke on the Water' comes to mind). Each line of iambic pentameter (the metre with which a lot of Shakespearean sonnets and verses are crafted) lasts for roughly three seconds when spoken. Segmentation of spontaneous speech acts belongs to this category too and so does the length of spontaneous motor activity (e.g., scratching or yawning) in different mammalian species. Information retrieving by short-time (working) memory is within three-second windows. The reproduction of time intervals is overestimated for time intervals smaller than three seconds, and underestimated for intervals greater than three seconds: the 'indifference point' therefore being at three seconds.

But can any quantum effects really last that long in the brain?

This is to look more macroscopically at the brain function. Zooming closer in, is Matthew Fisher and his brave journey into the nitty gritty of quantum mechanics working inside the brain, prompted by his personal desire to understand the unfathomable mental illness in people close to him. Fisher takes the baseline that there are lots of puzzling things about neuroscience, and asks whether there can be a better quantum explanation.

It has already been suggested that the protein tubes that make up the neuron's support structure could be in quantum superposition, meaning they could store twice the information of their classical counterpart. This has not been proven yet because qubit networks are too fragile and break if exposed to heat, mechanical interference etc., but Fisher would not stop at that.

'Life has had billions of years to "discover" quantum mechanics, and its exquisite molecular apparatus gives it the means to exploit it.' He starts with the spin as a quantifier of how much a nucleus 'feels' electric and magnetic fields — the higher the spin, the greater the interaction. A nucleus with the very lowest possible spin value, 1/2, feels virtually no interaction with electric fields and only a very small magnetic interaction. So in an environment such as the brain, nuclei with a spin of 1/2 would be peculiarly isolated from disturbance. Spin-1/2 nuclei are not common in nature: the spin value of lithium-6 (which is present in a lot of medication to treat symptoms of mental illness) is 1, but in the sort of chemical environment found in the brain, a water-based salt solution, the presence of the water's extra protons is known to make it act like a spin-1/2 nucleus. Experiments in the 1970s found that lithium-6 nuclei could hold their spin steady for up to 5 minutes. And so Fisher deduced that if there is an element of quantum control to the brain's computation, then lithium's calming effects might be down to the incorporation of these strangely coherent nuclei into the brain's chemistry. Although Lithium-6 does not occur naturally in the brain, one nucleus with a spin-1/2 does, and it is an active participant in many biochemical reactions. It is phosphorus — which led to Fisher concluding that 'if quantum processing is going on in the brain, phosphorus's nuclear spin is the only way it could occur.'

Fisher has taken into the public eye his 'candidate qubit' — a calcium phosphate structure known as a Posner molecule or cluster. If this produces the results expected, the brain's extracellular fluid could be found to be entwined with complex clusters of highly entangled Posner molecules. And when these are inside the neurons, these molecules could begin to alter the way the cells signal and respond, starting to form thoughts and memories. Exciting stuff. 'I'm still at the stage of telling stories,' the delightfully-modest Fisher tells us. 'I have to get some experiments done.' Another example of the act of desperation that proceeds all breakthroughs in science.

Quantum mechanics at work in the brain would explain a lot. Is this why a bang on the head induces memory loss, because it causes decoherence? Is nuclear spin the reason you can change brain states with transcranial magnetic stimulation that fires a magnetic field across the brain? How about Richard Alexander's views that the true motivations for human moral and social behaviour are evolutionarily designed to stay hidden from the consciousness? Could we think of the communication between the subconscious and conscious parts of our mind as a form of quantum communication? The subconscious part uses a form of quantum interference that, if accessed by the conscious part, would actually fail to give a desired outcome? This has an interesting parallel in quantum computing. Namely, how do we know that a quantum computer has executed the calculation that we asked it to do?

Add those to the even bigger questions — if the very stuff of our brain activity is quantum, what does that say for our thoughts, our actions, our behaviour on a large scale? For instance, do human thoughts come in discrete chunks we could call "thoughtons" in analogy with photons which are quanta of light? Perhaps the leap from social sciences to quantum physics is not so great after all. Maybe, these leaps are no greater than the exact distance we must go to provide the evidence that will bring these chillingly comprehensive speculations into solid, groundbreaking shapes.

EPILOGUE
The World, the Flesh and the Devil

The title of this closing chapter is that of a book written in 1929 by J. D. Bernal (another example of a non-religious scientist using religious language because of its higher emotive power). Considered one of the most visionary predictions ever made about the future of humanity, this book imagines three categories of obstacles that stand in the way of humanity's progress.

The first is the World — our environment, the planet and the rest of the universe. We need to fight the elements, guard ourselves against contingencies — such as asteroids hitting the earth — and make sure we do not run out of resources (such as fuel). Bernal conjectured that this is our least problem, because of the rapid technological progress that was already apparent at his time — but still a limitation.

The second class of obstacles is our biological make up, the Flesh. Our bodies are flexible, yes — amazingly resilient and even renewable, but they do have an expiry date written all over. Here Bernal's foresight becomes remarkable as he predicted that molecular biology would be able to tweak and improve this (considering that his book was written 30 years before the discovery of DNA).

And the last set of obstacles, which Bernal thought both the trickiest to predict and to tame. There are many different names for it — the ego, the uncontrolled mind — but it is our inner feelings, fears, desires and drives that make up the Devil.

In Christianity, the World, The Flesh and the Devil represent the forces of darkness standing in opposition to the Father, the Son and the Holy Ghost. Perhaps they are great barriers to human progression — but what about turning that around and using them (particularly the third) to aid our progression, or at least our understanding? If it is the human mind that created the gaps we have met with during this journey, perhaps it is also the human mind that can close them.

Science often gets a bad rep for asking more questions than it answers. Leo Tolstoy famously commented on what invariably happened when he asked scientists about the meaning of life: 'I received an innumerable quantity of exact replies concerning matters about which I had not asked.' Historically, of course, there was much less of a gap in our perception of nature. Religion, philosophy and science (which did not exist the way we think of it now) were inseparable, just as they frequently were for the Ancient Greeks and Romans. One might argue that this was good for the serenity of the mind. The world appeared as a whole — no gaps, no chasms in understanding or, what would be even more troubling, in the functioning of the universe.

Sadly, the very comforting nature of that picture was precisely because it was too simplistic. As our understanding grew and as technology advanced, we got our (increasingly powerful) magnifying glasses out. One by one, sciences separated from philosophy and religion, starting towards the end of the Renaissance. First physics, followed by chemistry and then finally, biology. The order in which they have become independent is not accidental as we talked about at the start of our journey: it follows the increase in complexity of things that we try to understand.

We have placed a high value on consistency when attempting to understand reality. Consistency means that our understanding will never contain statements that obviously lead to a contradiction. And yet humans can and clearly do hold contradictory views on many things. Many people

I know do not think twice about downloading pirated software and media over the internet, but they would not dream of stealing from a brick and mortar shopping mall. Most people say they appreciate the importance of a balanced life, but dedicate 75% of their life to work and an additional 10% to watching TV. People deny the possibility of gods outside of their own belief system, yet become very defensive when someone reciprocates that point of view.

And that does not make humans grind to a halt. On the contrary, some suggest that this is one of the defining features of being human. What if reality is the same? Could it be that the universe is inconsistent itself at the most fundamental level? This is certainly a logical possibility and has been entertained by many. Jorge Luis Borges, an Argentinian writer with a phenomenal mind for the bizarre, says: 'we (the undivided divinity operating within us) have dreamt the world. We have dreamt it as firm, mysterious, ubiquitous in space and durable in time; but in its architecture we have allowed tenuous and eternal devices of unreason which tell us it is false.'

George Orwell, English writer and social commentator, had a sinister take on this. He coined the word 'doublethink' in his novel 1984, to describe the state of having two contradictory thoughts in one's mind. He defines it as such: 'To know and not to know, to be conscious of complete truthfulness while telling carefully constructed lies, to hold simultaneously two opinions which cancelled out, knowing them to be contradictory and believing in both of them, to use logic against logic, to repudiate morality while laying claim to it (...) Even to understand the word "doublethink" involves the use of doublethink.'

His Gödelian ending here reminds me again of the shaving (or not shaving) barber paradox. Our picture of reality could therefore lead to a paradox at the heart of our understanding, and perhaps at the heart of reality itself. Like the Taoist I spoke to when climbing the Great Wall of China, who explained why a Taoist would not believe in the existence of the theory of everything, the one that would explain physics, chemistry, biology, economics and sociology all from a small set of equations. Because that which gives rise to everything, according to Taoism, has got to be infinite and, therefore, indescribable. This is mysticism at its best. The old infinity problem once again. The only way out of the infinity problem, as I see, is quantumly. Where would that fit in here?

In Plato's 7th book of *The Republic* is the well-known analogy of the cave (or the movie "The Truman Show" in modern parlance). Socrates, Plato's protagonist, invites us to see 'in a figure how far our nature is enlightened or unenlightened: — Behold! Human beings living in an underground cave, which has a mouth open towards the light.' A group of men have been in the cave since childhood, their legs and necks chained so that they can only see before them — they have only ever seen the shadows cast from a fire behind them. One man breaks the chains and escapes: at first, he is blinded by the real light of the sun, but soon he realises the greatness of the real world. He goes back to tell his fellow prisoners, but they are suspicious. Like Plato's prisoners, we see only shadows of appearances of real objects and hardly the actual fundamental nature of what is causing them. To grasp the real nature we need to free ourselves from our chains and leave the cave.

Physics, more than any other human activity, seems to me to be able to reveal the non-obvious aspect of the universe. And quantum physics, in particular, is not just the most accurate description of natural phenomena, but at the same time, is the most counterintuitive. It shows us more than any other pursuit that we are very akin to Plato's prisoners and require deep insight, arrived at after a long and laborious search, regarding what the ultimate underlying reality really is. Before we embarked on our voyage through the micro and macro, before we even left Oxford at the start, I outlined what I thought were the two greatest gifts we would gain if we were able to achieve the Great Reduction. One was technology, the other spirituality. Matter comes before the spirit, and I will talk about that now.

Basing our world view on just what our senses are telling us, we are naturally inclined to be materialistic. What you see is what you get. We perceive that the world is populated by stuff, such as tables, chairs, trees, rocks, animals and so on. All this stuff can be seen, touched, heard and felt. However, quantum physics teaches us that matter is mainly void and that its properties are best described by a mysterious function — the Psi function, of which we certainly have no direct experience. Schrödinger called Psi the catalogue of information, as it catalogues all we know about the system we are studying (be it an atom, or a molecule or a more complex solid such as a computer microchip). Psi tells us about all the possible things the system can actually do when we interact with it, but it does not specify — indeed the Heisenberg uncertainty principle prohibits us from this — exactly what will happen. The system will do so randomly, but according to the propensities specified by Psi.

Quantum physics is then all about how Psi changes in time. If we probe the system now, there will be one set of propensities: if we probe it in one hour, the propensities for different things would have changed. The law that describes these changes is called the Schrödinger equation. The important point is that neither Psi nor the Schrödinger equation give us any certainty about what will happen. All we can predict are the probabilities for various outcomes if the experiment is repeated many times.

So in quantum physics, the matter has actually become much less materialistic, much less actual, and becomes more like a cloud-like bundle of possibilities. We may find an electron somewhere, or an atom somewhere, but we can never know for sure until we measure. And after the measurement, the Schrödinger equation tells us that the system restores its state of uncertainty. Making a measurement in quantum physics establishes a definitive outcome and therefore, creates reality, but it is like making footprints in the sand. There is a sense of impermanence in quantum physics, where reality gets continuously created and destroyed, that makes it very difficult for us to remain materialists.

But quantum physics, by making the world less real, has actually simplified our reality and made a great unification in our understanding of the world. In particular, the distinction between energy (light) and matter has actually evaporated with the discovery of quantum physics. And this has allowed us to make inroads into chemistry and biology. I have also argued that physics is now starting to affect the social sciences too.

This point is frequently under-appreciated, but our technologies — that are currently getting more and more underpinned by quantum physics — are also moving further and further from materialism. We are moving towards integration of nano-technologies with our biological makeup. Nano robots, such as implanted mobile devices, will soon be part of our bodies and smart clothes will be able to regulate many things for us, such as our work, planning our trips and automating them. We could all have pocket-size quantum computers by 2050 and this will enormously improve our efficiency through the ability to do many things, most of them probably unimaginable from current perspective.

The integration is leading to greater internet connectivity and presence, to driverless cars and smart and connected houses and smart cities. And all these trends embody our departure further and further away from materialism. Consider this. The world's largest taxi firm, Uber, owns no cars. The world's most popular media company, Facebook, creates no content. The world's most valuable retailer, Alibaba, carries no stock. And the world's largest accommodation provider, Airbnb, owns no property.

This was spotted last year by Tom Goodwin, an executive at the French media group Havas, and has been doing the rounds since. His key point was that companies that control the interface between the consumer and the provider of the goods or services are in an incredibly valuable position. They carry none of the costs of providing the service, but take a cut from the millions of consumers who buy from them. The interface is where the profit is.

But the point that I am making here is that we are moving more and more into the digital arena where information (rather than matter) holds sway. In that sense, even our everyday world, that of human interaction, business, leisure, education and entertainment has become much less based on material things and much more on information. It is a revolution that transformed physics about 100 years ago and is currently starting to permeate all spheres of our life.

Ian Goldin, the Director of the Oxford Martin School, called me up one day and asked if I could recommend something to Richard Branson to take with him on his first Shuttle journey from London to Sydney. I was stunned. Richard Branson? For a start, I did not even realise that his first Shuttle flight was scheduled to happen so soon. Secondly, Branson is asking me, a quantum physicist, if I can give him something that would fit into a rucksack to take on board during his journey. Seriously? I mean, what could I possibly do for him?

My mind went blank in the way it does when it needs to work under pressure. Then I remembered my friend Alex Ling, another quantum physicist. Alex is my colleague in Singapore, and a master of miniaturising quantum experiments. You give him a bunch of bits of optics directing various laser beams from one end of the lab to another, and he will give you the same kit but shrunk down to 10 square centimetres. Like the dad in *Honey I Shrunk the Kids*, but slightly less mad. The best example of his genius is a compact photon pair source that he had embedded in a nanosatellite, designed to perform pathfinder experiments leading to global quantum communication networks using spacecraft. His equipment was sent up in a spacecraft launched in Germany: the fate of the spacecraft was not brilliant — it crashed, well, exploded, in Switzerland — but Alex's technology was found intact amongst the wreck. Post-recovery data showed no degradation in brightness or polarisation correlation. He proved that it is possible to engineer rugged quantum optical systems. That is what I am sending Branson, I thought.

In the quantum world, objects exist in many different places at the same time and their physical states can be more correlated than is classically allowed. These correlations (known as entanglement) allow us to achieve technological feats, such as teleportation, that are impossible classically. But to see these intrinsically quantum phenomena requires us to leave the imprisonment made by the cave of classical physics and use refined technology that allows us to interact with the quantum world. Transcending the world of classical physics has probably been the greatest triumph of the scientific method to date. The bonus of our impulse to leave the cave, of course, is the astounding development of our technology, which also contributes to our increased wellbeing (when used correctly and not abusively, of course). Perhaps this is the time when humanity will split into two: space explorers and back-to-basics environmentalists.

On small scales, quantum physics explains the structure of matter (and how the void that is dominant can still give rise to the solidity we feel), the laws of chemistry and through it, hopefully one day, it will also explain the laws of biology. How far this can go is by no means clear, but it has already gone quite a distance. There are even some claims that life is a necessity given the laws of physics.

It is plausible that the biggest scale structures, galaxies and their clusters, can too be explained through the laws of quantum physics. Some accounts of the universe are based on the quantum tunnelling of the universe into its own existence. These accounts are currently fully compliant with the existing astronomical observations. The small quantum jitters at this early stage of the universe are then amplified by a rapid expansion and are believed to lead to all the structure we see around us.

Quantum compression might also be able to explain the difference in the dynamics of entropy (which quantifies disorder in the universe) and complexity (which tells us how compressible the universe is). Namely, the universe is meant to start in a very ordered (low entropy state) and continue towards maximum entropy just as the second law of thermodynamics stipulates. For complexity, however, the universe starts in a simple state, and ends in a simple state. Sometime within these two extremes (now?) it is meant to reach a maximum complexity. It might not be surprising that creatures like us, capable of understanding the universe, arise during this most complex epoch. And understanding where we come from and why would, of course, bring a great deal of satisfaction. J. L. Austin said in 1979: 'It's not things, it's philosophers that are simple.'

Then we get on to reductions achieved by quantum field theory: merging quantum theory and special relativity has emphasised the 'undivided wholeness' (remember that the merger with general relativity is still outstanding); in relativity, this is because of the fact that there are no particles and rigid bodies in general; in quantum physics, this is because of the fundamentality of the field. It could give us a more accurate description of the evolution of the whole universe, and quantum field theory already dissolves boundaries between objects by treating them only as excitations of the field. Is this the key to bridging the gap or just an accident? And if it comes closer to closing the gap, is this gap still intrinsically unbridgeable? I am certainly not hiding from the fact that quantum physics cannot be unified with the general theory of relativity. Yet.

Jorge Luis Borges was an Argentinian writer of unusual, contemplative fiction. His stories are the literary equivalent of Esher's paintings, exploring the subjects of infinity and containing information about the whole universe in one point. They explore paradoxes such as Zeno's 'Achilles and the Turtle', the problem of infinite regression and the nature and reality of parallel worlds.

In another story, *The Library of Babel*, Borges imagines an infinite library that contains all possible books that could ever be written. A book that tells the exact story of your life (including the future and how you will die) is also in the library. However, there are two problems with it.

One is that it is impossible to find because a different book exists for every (infinite) potential of all your future actions; the other is that there are infinite other books which differ from it in some minute details (such as the exact time or the way you will die).

This story is another allegory on the Halting problem. Greg Chaitin's omega number is another, which brings me to the end of my conjectures on this topic. Is randomness the foundation of everything?

In the 1960s, Chaitin took up where Turing left off: fascinated by Turing's work, he began to investigate the Halting problem. He considered all the possible programs that Turing's hypothetical computer could run, and then looked for the probability that a program, chosen at random from all the possible programs, will halt. The work took him nearly 20 years, but he eventually showed that this 'Halting probability' turns Turing's question of whether a program halts into a real number, somewhere between 0 and 1.

Chaitin named this number Omega. And he showed that, just as there are no computable instructions for determining in advance whether a computer will halt, there are also no instructions for determining the digits of Omega. Omega is uncomputable. Some numbers, like *pi*, can be generated by a relatively short program which calculates its infinite number of digits one by one — how far you go is just a matter of time and resources. There is no such program for Omega: in binary, it consists of an unending, random string of 0s and 1s. The same process that led Turing to conclude that the Halting problem is undecidable also led Chaitin to the discovery of an unknowable number.

An unknowable number would not be a problem if it never reared its head. But once Chaitin had discovered Omega, he began to wonder whether it might have implications in the real world. So he decided to search mathematics for places where Omega might crop up.

And an obvious place is number theory, which is what Gödel used to demonstrate his incompleteness theorem. It describes how to deal with concepts such as counting, adding, and multiplying. Chaitin's search for Omega in number theory started with 'Diophantine equations' — which involve only the simple concepts of addition, multiplication and exponentiation of whole numbers.

Chaitin formulated a Diophantine equation that was 200 pages long and had 17,000 variables (ultimate dedication!). Given an equation like this, mathematicians would normally search for its solutions. There could be any number of answers: perhaps 10, 20, or even an infinite number of them. But Chaitin did not look for specific solutions, he simply looked to see whether there was a finite or an infinite number of them.

He did this because he knew it was the key to unearthing Omega. Other mathematicians had shown how to translate the operation of Turing's computer into a Diophantine equation, finding that there is a relationship between the solutions to the equation and the Halting problem for the machine's program. Specifically, if a particular program does not ever halt, a particular Diophantine equation will have no solution. In effect, the equations provide a bridge linking Turing's Halting problem — and thus Chaitin's Halting probability — with simple mathematical operations, such as the addition and multiplication of whole numbers.

Chaitin had arranged his equation so that there was one particular variable, a parameter which he called N, that provided the key to finding Omega. When he substituted numbers for N, analysis of the equation would provide the digits of Omega in binary. When he put 1 in place of N, he would ask whether there was a finite or infinite number of whole number solutions to the resulting equation. The answer gives the first digit of Omega: a finite number of solutions would make this digit 0, an infinite number of solutions would make it 1. Substituting 2 for N and asking the same question about the equation's solutions would give the second digit of Omega. Chaitin could, in theory, continue forever. 'My equation is constructed so that asking whether it has finitely or infinitely many solutions as you vary the parameter is the same as determining the bits of Omega,' he says.

But Chaitin already knew that each digit of Omega is random and independent. This could only mean one thing. Because finding out whether a Diophantine equation has a finite or infinite number of solutions generates these digits, each answer to the equation must therefore be unknowable and independent of every other answer. In other words, the randomness of the digits of Omega imposes limits on what can be known from number theory — the most elementary of mathematical fields. 'If randomness is even in something as basic as number theory, where else is it?' he asks. Chaitin is not one to leave a question unanswered. 'My hunch is it's everywhere. Randomness is the true foundation of mathematics.'

The fact that randomness is everywhere has deep consequences, says mathematician John Casti. It means that a few bits of mathematics may follow from each other, but for most mathematical situations those connections would not exist. And if you cannot make connections, you cannot solve or prove things. All a mathematician can do is aim to find the little bits of mathematics that do tie together. 'Chaitin's work shows that solvable problems are like a small island in a vast sea of undecidable propositions,' Casti says.

The fastest speed at which we might be able to flip a bit in this universe is the Planck constant divided by the total energy in the universe, and this is about 10 to the power of minus 50 seconds. Therefore, even if we started enumerating all the bits in the universe, we would have reached about 10 to the power of 70, which is 50 orders of magnitude smaller than the estimated capacity of the universe. Even if the universe, according to this logic, contains a finite amount of information, the actual speed at which we can uncover new information will never allow us to write it all down. So the Halting proof does not apply, yet the conclusion is the same: we can never exhaust all the bits in the universe.

The important point here is that the laws of physics determine what can and cannot be done. It does not matter if the laws of physics allow universal computation. They do, and all our computers testify to that. What matters is whether the universal computer as defined by Turing can actually simulate the laws of physics. Roger Penrose, the Oxford physicist we met in Chapter II, thinks that there are natural physical processes that cannot be simulated with computers.

Interestingly, the process that Penrose identifies as non-simulable lies at the boundary of quantum physics and gravity, where the gap in physics is the biggest. Maybe that is a logical place to expect to be non-computational. After all, we saw that it is possible that gravity has a profound effect on quantum physics and we are very much uncertain how to tackle this problem. Penrose claims that gravity makes quantum superpositions collapse.

In other words, the quantum property of being in many locations at the same time cannot be maintained under the influence of gravity. When this process decides where the system is ultimately localised, the way how this decision has been reached is not something a computer can simulate, according to Penrose. The Schrödinger cat is in this picture really dead or alive (and not both) and the reason is that gravity forces one of the outcomes. The process is, however, outside of computer powers to capture and model.

Sure, this is just Penrose's view. But provocatively, he uses this logic to challenge Turing's original motivation, namely to use machines to simulate human thinking and consciousness. What if quantum gravitational processes play a role in the human brain too? If they determine whether the cat is dead or alive, they also could influence if we are happy or sad (and not both). That would also mean that computers are incapable of simulating the process of human thinking.

Good news for science and human creativity! While computers are incapable of solving the Halting problem, maybe humans can. This might mean that we can, despite the Halting issues, bridge the gap between physics and chemistry, and physics and biology, albeit not using computational simulations. Inspiration, insight and other aspects of consciousness could be the key in bridging the gap.

Allow my speculation: we do not know if quantum physics when merged with gravity will lead to anything really different to what we are used to. We could get a new theory of physics, yes, but this might not imply that there is something incomputable about it. The point is not that this logic fails, it is that in principle, the Turing thesis could just not be right. The laws of physics could allow something to happen that is not computable. And that simply means that we cannot use the Halting problem to claim that the micro-macro gap will always remain open.

We have already had a little look at the technological advantages that would come about if the micro-macro gap is closed. How about our spiritual yearnings — the desire for connection, completion, and understanding of our 'higher' or inner, purpose — whatever your words for it are, I believe it is important to our well-being. And individual well-being is the only thing that can make societal well-being possible. It would mean that these seemingly different benefits, technological and spiritual, are not so separate. It is that I want to discuss now, within the main context of this book. Would a bridge between our micro and macro understanding of reality enrich us spiritually?

The pursuit of happiness is of such paramount importance to us that the founding fathers of the US felt its need to be protected by the constitution and so listed it as one of the self-evident truths. We might follow different paths on this road to happiness, yet most of the things we search for are common to us all. Some of these are obviously of a great evolutionary value, such as cultivating the land, or having children and finding a suitable environment in which to raise them, but others are more subtle and of a less tangible nature. Of course, a large proportion of our population is still struggling at subsistence level, and do not have the time or means to contemplate their place in the universe. But something that seems shared among most people, irrespective of their material situation, is the yearning for meaning. The main difference is that it is something the materially more fortunate might be able to devote larger fractions of their time towards searching for.

We very often yearn to understand our origins — the origin of life, the origin of the universe, and the meaning of it all. None of these are per se of any immediate biological value (as the existence of other living creatures illustrates) and yet many people spend a lot of their time and mental energy fixed on these questions. After all, according to The Doors (what better authority than Jim Morrison): 'Into this house we're born, into this world we're thrown' — and we devote a substantial amount of our efforts to understanding the meaning of it all. Why?

Trying to understand who we really are seems to make us happy. Is it the sense of control? Or feeling of transcendence? Perhaps it helps to ground us into the world around us. It is as if something went out of sync during humanity's evolution, a piece of the puzzle that is missing, that we often want desperately to fill in. The gap between micro and macro, self and universe.

Two things seem to me common to humanity's search for bridging that gap, whether through science, music, philosophy, religion, or a million other things. Things that both still the mind enough from its daily chatter to engage with the 'bigger picture'. One is this feeling of transcendence: that we often feel, at some point in life, that there is more to the universe than meets the eye. Something that goes beyond what our limited senses can see, hear, touch, smell or feel. And this ' beyond' is accessible to us by seeking to close the gaps in search for deeper understanding.

Another of Borges' stories, *The Aleph*, is about a point in space (called the Aleph) that contains all other points. Anyone who gazes into it can see everything in the universe from every angle simultaneously, without distortion, overlapping or confusion. An attractive idea. The Aleph is in the basement of Carlos Argentino Daneri, a mediocre poet with a vastly exaggerated view of his own talent, who has made it his lifelong quest to write an epic poem that describes every single location on the planet in excruciatingly fine detail. The narrator, of course, wants to see it for himself. Who would not?

Transcendence is the macro commonality of our experience of this gap, and the other is much smaller and closer to earth: the order and simplicity of our natural world. It is actually remarkable how far Occam's razor, 'assumptions should not be multiplied beyond necessity', can be taken when trying to understand natural phenomena. It is simply mind-blowing how much actually follows from a handful of physical principles. Our feelings of transcendence at a large scale, and wonder at a micro one, are all too often blocked by the incessant chatter of our minds. I cannot put it more beautifully than Blake: 'If the doors of perception were cleansed, everything would be seen as it is, infinite'. Plato called this higher reality the world of ideas in which perfect mathematical forms exist in a timeless fashion and are only imperfectly imitated by the physical world we inhabit. Every chair in this world only resembles the (Platonic) perfect chair existing in the sphere of ideas.

This is echoed in many religions. Jesus is for Christians the point of contact between the world of God (i.e. the Platonic one) and the world of man. Looking further east, Hinduism and the Buddhism that emerged from it centre around transcendence: their central practices are routes to higher ways of being through the state of meditation, meant to elevate us from our everyday existence and help us get in touch with the other, transcendental, deeper mode of existence. What is this block that stands between our physical world and Plato's 'higher reality'? Our concept of time, I believe, has a lot to answer for here.

I am thinking of the poor White Rabbit in *Alice's Adventures in Wonderland*, brought to utter distraction by his watch and insistent 'I'm late!'s. The fading away of time as we go deeper into the micro domain as well as its central importance in the macro domain suggest to me that time will be crucial to understanding the two. Furthermore, there are good reasons to suppose that at the very largest scales of the universe as a whole, time again disappears. Our understanding of time strongly impacts how we see ourselves and our place in nature. One of the curious differences between the micro and macro worlds in physics is that time features prominently in the latter, but it is strangely evasive in the former.

What is time? There are many different aspects of time and many different types of time. Its most prominent feature, however, is its flow, which is responsible for the fact that everything that comes into existence, then develops, changes and ultimately dies. The creative and destructive aspects of time are both a blessing and a curse.

In physics, we are very pragmatic towards time. It is a parameter that measures the rate of change of things we observe. Curiously, time itself is never measured directly. We use the periodic motion of stars, planets, pendula, and, more recently, atoms, to track time. Time, in other words, is always measured using the position of some object (be in a star, planet, pendulum, or the electron inside an atom). When I measure how quickly I run, I actually look at my position relative to the position of the hands of my watch. All time is, according to this, is the relative position between two objects (in this case, myself and the hands of my watch).

This is rather curious since for a watch to keep time accurately for a sufficiently long time, it needs to be sufficiently large. It has to have enough capacity to record time as well as not to drift too much during its recording. It has taken us a long time to perfect our time-keepers, but at present our best watches would lose only one second during the evolution of the whole universe (13.7 billion years, or about 10 to the power of seventeen seconds). Good work, humanity.

How about the level of nuclear and sub-nuclear particles? How long can a nucleus keep track of time? The answer is about a millisecond and here is how this comes about. Currently, we think of space in tiny units known as Planck's length (ten to the power of minus 35 of a meter) and time in terms of tiny ticks known as Planck's time (an equally stupendously small ten to the power of minus 43 of a second). A nucleus is about ten to the power of minus 15 meters, which gives us the capacity of about ten to the power of 40 bits. Therefore, a nucleus could keep track of 10^{-3} seconds, which is a millisecond. And that is really nothing.

For smaller object still, time becomes more and more meaningless, until we arrive at the scales where it is actually meaningless to talk about it. There is a strong suspicion in the physics community that time is not fundamental, but emergent. It starts to make sense only for objects like us, that are of sufficient complexity. In other words, time is a macroscopic construct.

There is a sense in which the problem of quantising gravity, of the physics micro and macro, is actually mainly a problem with understanding time.

The physicist W. G. Unruh said "Gravity is the unequable flow of time from place to place.'. In particular, time (as measured by any clock) actually runs slower closer to a massive object. If you stand on top of a mountain, in other words in a place where gravity is weaker than at the bottom, you will actually age faster. The difference, even if you have climbed to the top of the Mount Everest, is so small that it would not make much effect on your biological existence (it will shorten it by a tiny fraction), though an atomic clock will definitely tick noticeably differently to the one at the bottom (since atomic clocks can measure a billionth of a millionth of a second).

Gravity is now seen as the law saying that objects move so that their time runs fastest given other conditions are fixed. Suppose we throw something in the air. The object first goes quickly up, then slows down coming to a stop at a certain height (depending on how hard we have thrown it). The object then starts to slowly accelerate back down, picking up speed and arriving quickly back at the original place. Einstein's general relativity — our best theory of gravity — explains that this is because the clock wants to spend as much time higher up as possible because this is where the time runs faster!

If we throw the object at an angle, it will result in a motion describing a parabola, again a consequence of wanting to age as quickly as possible, according to Einstein.

The Psi function in quantum physics, the catalogue of information telling us the propensities for different future behaviour, also changes more quickly if the object is higher. However, there is no consistent way of calculating how this really ought to work. Gravity is all about time, while quantum physics says that time does not really exist; it is never really measured directly, only by looking at something else (like hands of a clock). These two different pictures of time clash when we try to quantise gravity, always giving us nonsensical answers.

In fact, the micro and the macro pictures also permeate the artistic visions of the world. Walter Benjamin, a German Jewish cultural critic, compares Proust and Michelangelo in the following manner. Proust's writing style is life as a sequence of unconnected snapshots. Every instant of time is self-contained and independent from the greater whole (there is no such thing as far as his novels are concerned). Michelangelo, on the other hand, offers us a great single vision of the whole reality in his painting on the ceiling of the Sistine Chapel. Both Proust and Michelangelo offer a cosmic image, but at the opposite poles of time: Michelangelo gives us the world as macrocosm; Proust as a network of microcosms.

At the core of many religions lies the concept that the passage of time and our ephemeral existence are at the root of all suffering and discord. The way of overcoming the pain of our short existence is to make contact with the timeless domain, in which things just are — they do not change, but exist once and for all. I would like to speculate that the closing of the micro and macro gap could also present a spiritual transformation for humanity much in accordance with these ancient doctrines. This will come with the realisation that time exists only at the intermediate level and is completely absent for the microscopic systems as well as the universe as a whole. The small systems do not have enough capacity to keep track of time for long enough, whereas the universe as a whole just is — there is no external clock to the universe with respect to which to measure time. 'The essence of nowness,' Santayana says, 'runs like fire along the fuse of time.'

It is curious that physics, while trying to bridge the quantum and gravity micro-macro divide, has actually enabled us to understand that there is no contradiction between a timeless universe at the grandest scale and the flow and arrow of time we acutely feel at our own level of existence. English philosopher Robert Hooke says 'I would query by what sense it is we come to be informed of Time; for all the information we have from the senses are momentary, and only last during the Impressions made by the object. There is therefore yet wanting a Sense to apprehend Time.'

Along with suggestions that time actually emerges with life, it follows that its linear character actually has to do with data compression in our memory. By linear character, I mean that there is (apparently) only one dimension of time. We take it as given that while there are three dimensions of space (up-down, left-right and forward-backward), we can only move forward in time, one second per second and there is no room to manoeuvre beyond this. And data compression is a device that the g-zip programs in all computers use to reduce the size of your file while still retaining all the key information (so that you can always recover the original file by g-unzipping it).

Hooke concludes thus: 'I say, we shall find a Necessity of supposing some other Organ to apprehend the Impression that is made by Time. And this I conceive to be no other than that which we generally call Memory, which Memory I suppose to be as much an Organ as the Eye, Ear or Nose, and to have its Situation somewhere near the Place where the Nerves from the other Senses concur and meet.' What is the connection between the finiteness of memory and the perception of time and its flow?

Imagine that reality contains all the things that can actually happen. Let us say that in the next 10 instances of time we are aware of, there are 10 different things that can happen at each of these instances. The number of possible strings of event is then 10 to the power of 10. If we need to have the mental capacity to record this, then we need the same number of mental states in our memory. Interestingly, this is the number of possible synaptic bits we are estimated to have in our brain.

However, we need to record many more instances over our life-times (even if we record once every tenth of a second and we record only ten different possibilities this is ten to the power of ten billion!). This is clearly impossible, which is why memory needs to use tricks (like g-zip) to compress. How does the memory achieve this? My favourite suggestion is that memory (in order to handle this) uses the trick used by magicians to memorise cards.

Humans are bad at memorising random numbers, or images — but we are excellent at remembering stories. We love gossip. Our connections depend on it. What has been wrong with him or her this week? Where did so and so go on holiday? Who is dating whom? Who is sitting on our promotion committee? And so on. We remember whole stories as long as they are interesting intrigues related to human affairs. The best way to memorise cards is to convert their order into a story. For instance, a queen meets a jack at 10 minutes past 6, and they go to see the king and so on...this way you have memorised queen, jack, 10, 6 and king and with a bit of practice you will be able to do all 52 cards.

Can we remember the suits of the cards too? Sure. A simple scheme could be to think of diamonds as rich people, hearts as people you love, clubs as tough people and spades as amusing or absurd people. There are of course many other ways, just as there are many computer programs for data compression. The key feature of any story is that there is a point to it. There is a causal order. It makes sense for the queen and jack to meet. It makes sense that they go and see the king. The more we connect the story logically, the easier it will be to memorise and recall it at will.

The point is that if we use no tricks we need 52 times 51 times 50 times 49 and so on bits of memory to memorise a pack. But this is a humongous number far exceeding the estimated brain capacity. But by making a story out of it in which one event (card) naturally follows another, all we need to remember is a one dimensional sequence of things. This reduction for a universe of all possibilities to just one dimension seems reminiscent of perceiving only one dimension of time. Heinz von Foerster, an Austrian American scientist, says 'The conceptual construct of time is, as far as I can see, just a by-product of our memory, which in some instances might use time as a convenient parameter'.

One philosopher that has recently tackled the problem of the existence of time in a timeless universe is J. Ismael. Here she says: 'The reason that physics has done a decent job accommodating asymmetry, but not such a good job with flow, passage, and openness, I would suggest, is that asymmetry is an artefact of the shift from a microscopic to a macroscopic perspective, whereas flow, passage, and openness arise in the transformations wrought in that horizontal dimension.'

Adding the horizontal dimension allows us to close the circle, bringing experience and ontology back together as part of a single, unified vision of the universe. 'The wise focus on common interests', Confucius said, 'the unwise on differences'. We feel awe at the existence of order in the universe, but this can also lead to the uncomfortable fear of determinism. If physical facts fix all the facts in the world, does that not mean that all our actions are predetermined?

This, of course, is a deep question to which science has no answer at present. Suffice it to say that quantum physics certainly destroys Leibniz's 'principle of sufficient reason', according to which, everything that happens must have an underlying cause. The most elementary events in quantum physics are, as far as we know, intrinsically random and happen without any prior reason. However, whilst this might help us alleviate our fear of determinism, it is certainly not enough to guarantee any free will.

While I am personally happy enough to live with that state of affairs, I am also confident that we will discover more on this very issue in the years to come.

'It is sometimes said that scientists are unromantic, that their passion to figure out robs the world of beauty and mystery. But is it not stirring to understand how the world actually works — that white light is made of colours, that colour is the way we perceive the wavelengths of light, that transparent air reflects light, that in so doing, it discriminates among the waves, and that the sky is blue for the same reason that the sunset is red? It does no harm to the romance of the sunset to know a little bit about it.' To paraphrase Sagan, any new understanding can only add to the beauty of the world.

It is said that at the entrance to Plato's academy, the world's first university in the modern sense of the word, it was inscribed: 'Do not enter if you do not understand geometry'. After two and a half millennia, I feel we can afford to be a bit presumptuous as to suggest a small correction to the great Plato: 'Do not enter if you do not understand quantum physics'. As far as I am concerned, it is the road to all emotional and intellectual fulfilment.

ACKNOWLEDGEMENTS

There are many people who should be thanked and without whom this book would not have existed. I have been influenced by so many colleagues and friends in the past 20 years of my research career. In alphabetical order, those who did most are Charles Bennett, Keith Burnett, David Deutsch, Artur Ekert, Peter Knight, Chiara Marletto, William Wootters, and Anton Zeilinger. They will find much of themselves in the pages of this book.

Kate Sudwicks' masterful line editing and encouragement throughout is gratefully acknowledged. She provided much of the stimulus at the final critical stages of drafting and, without her help, this project would never have materialised.

REFERENCES

1. *Physics and Politics,* Walter Bagehot, (Cosimo Classics 2007).

2. *Quantum Physics: Illusion or Reality?,* Alastair Rae (Canto Classics 2012).

3. *The Ghost in the Atom,* Paul Davies (Canto 2010).

4. *What is Life?,* Erwin Schrödinger (Canto Classics 2012).

5. *The Selfish Gene,* Richard Dawkins (Oxford Landmark Science 2016).

6. *The Evolution of Cooperation,* Robert Axelrod (Penguin Press 1990).

7. *The World, the Flesh and the Devil,* J. D. Bernal (Indiana University Press 1969).

8. *Demons, Engines and the Second Law,* Charles Bennett, *Scientific American,* November 1987.

9. *Decoding Reality,* Vlatko Vedral (Oxford University Press 2010).

10. *Man and His Universe,* John Langdon-Davies (Thinkers Library No 61 1937).

11. *The Republic,* Plato (Classic Books 2017).

12. *The Art of Thinking Clearly,* Rolf Dobelli (Sceptre 2014).

13. *What is Life? How Chemistry Becomes Biology,* Addy Pross (Oxford University Press 2012).

14. *Time's Arrow and Archimedes' Point,* Huw Price (Oxford University Press 1996).

15. *Collection of Essays,* G. Chaitin (World Scientific, 2007).

16. *"Chance and Necessity",* Jacques Monod (Vintage 1971).

17. *Water's Quantum Weirdness Makes Life Possible,* Lisa Grossman, *New Scientist,* 19 October 2011.

18. *"Negentropy Principle of Information",* Leon Brillouin, *J. of Applied Physics,* **24,** 1152, 1953.

19. *Proton Tunneling in DNA and its Biological Implications,* P.-O. Löwdin, *Rev. Mod. Phys.,* **35,** 1963.

20. *Magnetic Compass of European Robins,* R. & W. Wiltschko, Science, **176,** 62, 1972.

21. *Physical chemistry: Quantum mechanics for plants,* G. R. Fleming and G. D. Scholes, *Nature,* **431,** 256, 2004.

22. *Toward Quantum Simulations of Biological Information Flow,* R. Dorner, J. Goold, V. Vedral, *INTERFACE FOCUS* Volume: **2** Issue: **4,** 522, 2012.

23. *Games of Life,* K. Sigmund (Oxford University Press 1993).

24. *Mathematical Theory of Communications,* C. E. Shannon and W. Weaver (University of Illinois Press 1948).

25. *The Stream of Life,* Julian Huxley (1928).

26. *Wholeness and the Implicate Order,* David Bohm (Ark paperbacks 1980).

27. *The Genetical Evolution of Social Behaviour. I,* Hamilton, W. D., *Journal of Theoretical Biology* 7 (1): 1–16, 1964.

28. *On Growth and Form*, D'Arcy Wentworth Thompson (Cambridge University Press 1917).

29. *Vicious Circles and Infinity*, Patrick Hughes and George Brecht (Jonathan Cape 1973).

30. *Perennial Philosophy*, Aldous Huxley (Harper and Brothers 1945).

31. *The Myths of Time*, Hugh Rayment-Pickard (Darton, Longman and Todd 2004).

32. *Temporal Experience*, J. Ismael, *The Oxford Handbook of Philosophy of Time* (Oxford University Press 2010).

33. *The Road to Serfdom*, Friedrich Hayek (University of Chicago Press 1994).

34. *The Prince*, Niccolò Machiavelli (University of Chicago Press 1985).

35. *The Art of War*, Sun Tzu (Oxford University Press 1963).

36. *Thinking, Fast and Slow*, Daniel Kahneman (Farrar, Straus and Giroux 2011).

37. *Instant Expert 33: Quantum Information*, Vlatko Vedral, *New Scientist*, 2013.

38. *Living in a Quantum World*, Vlatko Vedral, *Scientific American*, 2011.

INDEX

Index

Index

Printed in the United States
By Bookmasters